Dissertations in American Economic History

This is a volume in the Arno Press collection

Dissertations in American Economic History

Advisory Editor
Stuart Bruchey

Research Associate
Eleanor Bruchey

See last pages of this volume for a complete list of titles.

THE LOCATION OF THE UNITED STATES STEEL INDUSTRY 1879-1919

Ann K. Harper

ARNO PRESS

A New York Times Company

New York / 1977

Editorial Supervision: LUCILLE MAIORCA

First publication in book form, Arno Press, 1977

DISSERTATIONS IN AMERICAN ECONOMIC HISTORY
ISBN for complete set: 0-405-09900-2
See last pages of this volume for titles.

Manufactured in the United States of America

Library of Congress Cataloging in Publication Data

Harper, Ann K
The location of the United States steel industry, 1879-1919.

(Dissertations in American economic history)
Originally presented as the author's thesis, Johns Hopkins, 1976.
Bibliography: p.
1. Steel industry and trade--United States--History. 2. Industries, Location of--United States -- History--Mathematical models. I. Title.
II. Series.
HD9515.H33 1977 338.6'042 76-39829
ISBN 0-405-09909-6

The Location of the United States Steel Industry, 1879-1919

by

Ann K. Harper

A dissertation submitted to The Johns Hopkins University in conformity with the requirements for the degree of Doctor of Philosophy.

Baltimore, Maryland

1976

PREFACE

Many people have assisted me, directly or indirectly, on this project; I feel very indebted to them. Among the indirect helpers is Dr. Carl Stern of Randolph-Macon Woman's College; his enthusiasm for the study of economics was (and undoubtedly still is) quite infectious. For their more direct assistance, I would like to thank Dr. Bruce Hamilton and Dr. Peter Newman of Johns Hopkins University. Their keen criticism and helpful suggestions aided immeasurably in progress on the dissertation (had I followed all their advice more diligently, it would be a vastly improved product). Two colleagues at Western Maryland College, Professor Ethan Seidel and Dr. Ralph Price, made many useful comments on various portions of the text. Ms. Lola Mathias not only ably typed the manuscript, but also performed many duties beyond those required of a typist; further, she also maintained a reassuring cheerfulness which aided tremendously in the hectic moments of assembling the manuscript. Ms. Peggy Cermack and Mr. John Norment provided indispensable assistance with various computer programs. Professor Roy Fender contributed the graphs for the text. Finally, I would like to thank Meredith and Roy for their patience over the past few years.

TABLE OF CONTENTS

	Page
Chapter One -- INTRODUCTION	1
Chapter Two -- REVIEW OF EXISTING FIRM AND INDUSTRIAL LOCATION THEORY	6
Chapter Three -- DESCRIPTION OF THE STEEL INDUSTRY IN THE UNITED STATES, 1879-1919	68
I. Introduction	68
II. Technology	69
III. Processing of Steel into Shapes	81
IV. Raw Materials	83
V. Demand	92
A. Products of Steelworks and Rolling Mills	92
B. Geographic Distribution of Demand	97
C. Cyclical Fluctuations in Aggregate Steel Output	112
VI. Industrial Concentration in the Steel Industry	116
VII. Location of the American Steel Industry, 1879-1919.	130
VIII. Steel As A Weber Transport Oriented Industry	142
Chapter Four -- THE LINEAR PROGRAMMING MODEL FOR THE WEBER PROBLEM	145

TABLE OF CONTENTS CONTINUED

	Page
I. Introduction	145
II. Linear Programming and the Transport Oriented Industry	150
III. An Illustration	156
A. The Minimum Transport Cost Problem	156
B. The Dual	158
IV. Application of the Model	160
A. Computations and Data Sources	160
B. Results Generated by the Model	163
C. Optimal Shipment Patterns, Given the Actual Location of Production	173
Chapter Five -- INTERPRETATION OF RESULTS	187
I. Introduction	187
II. Statistical Analysis of the Divergence Between Optimal and Actual Output	190
A. Nonparametric Tests	190
B. Parametric Tests	200
III. Descriptive Analysis of the Divergence Between Optimal and Actual Locations	216
IV. Conclusions	235
Appendix A: Data Used in the Solution of the Model	245
I. The Problem	245

TABLE OF CONTENTS CONTINUED

		Page
II.	Transport Cost Data	245
III.	Input-Output Coefficients for Alabama	250
Appendix B:	Shipments of Coal, Ore in the Capacity Constrained Model	252
REFERENCES		258

TABLE OF ILLUSTRATIONS

		Page
Figure 1.	Locational Triangle	11
Figure 2.	Individual Demand Curve from Lösch's Locational Analysis	44
Figure 3.	Demand Cone from Lösch's Locational Analysis	45
Table 3.1.	Production of Iron and Steel, 1879-1919	70
Table 3.2.	Steel Production, by Kinds	78
Table 3.3.	Average Tonnage of Ore and Coal Per Ton of Steelworks' Product, 1879-1919	91
Table 3.4.	Products of the American Steel Industry, 1879-1919	94
Table 3.5.	Demand for Steel, 1879-1919	100
Table 3.6.	Geographical Distribution of Demand for the Products of Steelworks and Rolling Mills, 1879-1919, Assuming Growth in Each Category Accounts for 80% of Total Demand, and Absolute Size of Proxy Accounts for 20% of the Total	109
Table 3.7.	Geographical Distribution of Demand for the Products of Steelworks and Rolling Mills, 1879-1919, Assuming Decadal Growth in Proxy Variable for Each Category Reflects 50% of Demand in That Demand Category and the Absolute Size of Proxy Reflects 50% of Demand in That Category	110

TABLE OF ILLUSTRATIONS CONTINUED

		Page
Table 3.8.	Geographical Distribution of Demand for the Products of Steelworks and Rolling Mills, 1879-1919, Assuming Decadal Growth in Proxy Variable for Each Demand Category Reflects 20% of Demand in That Category and the Absolute Size of the Proxy Reflects 80% of Demand in That Category	111
Table 3.9.	Concentration of U. S. Crude Steel Production	128
Table 3.10.	Percentage of U. S. Steel Production By States	133
Table 3.11.	Share of Output as a Function of Time	140
Table 4.1.	Optimal Production of Steel and Shipments of Coal, Ore	165
Table 4.2.	Actual Output, Steelworks and Rolling Mills, 1879-1919	167
Table 4.3.	Production (in tons) of Steel Under the Minimum Transport Cost Model with Birmingham Input Requirements Adjusted	172
Table 4.4.	Pattern of Steel Shipments Generated by the Capacity Constrained Program, 1879-1919	178
Table 5.1.	Rank and Sign of Differences Between Optimal and Actual Output Levels and T Scores	193
Table 5.2.	Rank and Sign of Decadal Differences in Optimal and Actual Output Levels and Value of T for Each Test; Rank and Sign of Differences in Optimal and Actual Output Levels and Value of T, 1879-1919	195

TABLE OF ILLUSTRATIONS CONTINUED

		Page
Table 5.3.	Calculated Values of K_D and Critical Values of K_D at the .05 Significance Level	199
Table 5.4.	Regression Equations for Actual and Program Generated Output As Functions of Transport Costs	203
Table 5.5.	Significance at .05 Confidence Level of the Difference Between Coefficients of Actual Output Regressions and Optimal Output Regressions	208
Table 5.6.	Regression Equations for Actual and Program Generated Output As Functions of Transport Costs - Combined Sample Data	212
Table 5.7.	Significance at .05 Confidence Level of the Difference Between Coefficients of Actual Output Regressions and Optimal Output Regressions - Combined Sample	213
Table 5.8.	Program Generated Levels of Output by Location	222
Table 5.9.	Capital Costs Per Ton of Steel Output	225
Table 5.10.	Dollar Capital Required to Produce Program Generated Levels of Output	226
Table 5.11.	Excess Capacity (Dollar Value)	229
Table 5.12.	Transport Costs: Net Saving As a Result of Optimal Location	230

TABLE OF ILLUSTRATIONS CONTINUED

		Page
Table 5.13.	Transport Cost Reduction Due to Optimal Location As a Fraction of Estimated Actual Transport Costs	239
Table A-1	Input-Output Coefficients For Alabama	251
Table B-1	Raw Materials Shipments Under the Constrained Capacity Model	253

CHAPTER 1: INTRODUCTION

The late nineteenth century witnessed rapid growth in output and changes in technology, industrial structure, and location of many industries in the United States. This transitional period has long fascinated economic historians; a typical description of the post-Civil War era is provided by Kroos and Gilbert:

> Production and business activity soared as never before. Although many of the leading industries of the pre-1850 period continued to grow with the country, it was the appearance of new industries and new types of entrepreneurship that provided the romance and excitement of the nineteenth century. There emerged a group of business leaders patterned after Schumpeter's "heroic entrepreneur", although they were described by others in somewhat less flattering phrases. Be that as it may, by the end of the century, the United States had emerged as the industrial leader of the world.[1]

The steel industry exhibited many features characteristic of the period. Between 1879 and 1919 steel output increased more than tenfold; two

---

[1]Hermann E. Kroos and Charles Gilbert, American Business History, (Englewood Cliffs, New Jersey: Prentice-Hall, Inc.; 1972), p. 144.

new methods for producing steel cheaply, the Bessemer and open-hearth techniques, revolutionized steel production in the post-bellum period; and the industry, which had been characterized by more or less vigorous competition (and unsuccessful price fixing pools) after the Civil War, became dominated by the giant United States Steel Corporation after its formation by merger in 1901. This dissertation deals primarily with a less frequently examined aspect of post-bellum industrial development, the location of industry, in particular, the steel industry in the United States between 1879 and 1919.

This industry was an active participant, as a supplier of raw materials crucial in the building of private and public capital goods, in the industrial expansion of the United States between 1879 and 1919. That alone would make a study of the industry's location important to an overall understanding of American industrial development and location. Additionally, the iron and steel industry, with its weight losing inputs and the bulky nature of its product, is a frequently cited example of an industry whose location is very sensitive to transportation costs.[1]

The particular hypothesis to be examined here is that the steel industry was a Weber transport oriented industry; that is, the industry located so as to minimize transport costs. A linear programming model which seeks minimum transportation cost location is devised to test the hypothesis. Historical data are used in the model to generate minimum transport cost ("optimal") production locations for the steel industry in each of the census years 1879, 1889, 1899, 1909, and 1919. The optimal geographical distribution of output so obtained is then compared with the actual locational distribution of output. The linear programming model is also used to determine the minimum transport cost pattern of materials and steel shipments, given the actual location of production in each of the five census years involved.

---

[1]See, for example, Alfred Weber, Theory of the Location of Industries, trans., ed., and introduction by Carl J. Friedrich, (Chicago: University of Chicago Press; 1929). Weber outlined the conditions under which transport cost minimization was consistent with profit maximization.

For this study, the term "steel industry" refers to steelworks and rolling mills, a vertically integrated unit which produced semifinished rolled steel. In some years it is difficult to separate steel from iron production in the historical data sources, but the attempt has been made to distinguish the two, and to deal with steel production. In general, data sources for this period are limited, both in information and in consistency. Both extensive searching for data and extensions of the available information to fill gaps were necessary to obtain a numerical solution to the linear programming model.

Before detailing the location of the steel industry, Chapter 2 surveys that portion of location theory relevant to location of firms and industries. The third chapter provides background information on the turn of the century American steel industry: its technology, raw materials requirements, location, industrial structure, and the demand for its products. Chapter 4 explains the linear programming model which is solved for the minimum transport cost geographic distribution of steel production. The same chapter summarizes the data sources used to generate the numbers necessary for solving the model, and also outlines the results of the program. In the fifth chapter, the optimal (minimum transport cost) and actual locations of the American steel industry, 1879-1919, are compared. Finally, the usefulness of the linear programming technique for examining industrial location in an historical context is discussed. An Appendix contains a summary of the data generated for the study.

CHAPTER 2: REVIEW OF EXISTING FIRM AND INDUSTRIAL LOCATION THEORY

Although the locational study of the steel industry presented in the next few chapters is based on Alfred Weber's analysis of industrial location, the contributions of several other location theorists are surveyed, along with Weber's in this chapter. The advantages and limitations of the Weber approach become more clearcut when his approach is contrasted with that of other major contributors to the field. These major contributors include, in addition to Weber, Edgar Hoover, August Lösch, Harold Hotelling, Arthur Smithies, Walter Isard, and Martin Beckmann. Although others have written extensively in the field, and some of their names will be mentioned in passing, this review concentrates on the work of the above.

Alfred Weber was among the first to study industrial location. Weber emphasized supply factors rather than a more general theoretical framework in his major work on industrial location, translated as <u>Theory of the Location of Industries</u>.[1] He mentioned general factors in location, applicable to all industries, such as labor costs, transportation costs and rent. He distinguised locational factors of the regional type from agglomerative factors; the regional factors determined the extent to which firms cluster within a given region; and agglomerative factors generally entailed economies external to the industry. Weber treated these agglomerative factors, which could have negative as well as positive effects, as a single, net force. Since he considered the compositions of this force to be of little importance, Weber emphasized regional factors.

Weber suggested three regional influences on location: the relative price range of materials deposits, the cost of labor, and the cost of transportation. By expressing the relative price range of materials deposits in terms of transport cost difference, e.g., treating the cheaper materials site as though it were closer to the plant, Weber was left with only two regional factors. The two,

---

[1]Alfred Weber, <u>Theory of the Location of Industries</u>, trans., ed. and introduction by Carl Friedrich, (Chicago: University of Chicago Press; 1929).

transportation costs and labor costs, formed the basis of his analysis, with the net positive or negative agglomerative force forming a third locational force.[1] Edgar Hoover criticized this treatment of cheap materials deposits because it implies the plant site is known initially, rather than to be determined by solution of the locational problem.[2]

Weber was concerned with the interrelationships among these factors. To isolate them he assumed a given geographical distribution of materials, given location(s) of consumption, and given labor locations, each having an infinitely elastic supply of labor at the going wage and wages possibly varying in different labor sites.[3] Weber recognized somewhat the limitations these assumptions put on his analysis. Certainly the assumptions significantly reduced the possible interrelationships he could examine, and also made the analysis supply-oriented.

---

[1]Ibid., pp. 34-5.

[2]Edgar Hoover, Location Theory and the Shoe and Leather Industries, (Cambridge, Massachusetts: Harvard University Press, 1937), p. 36.

[3]Weber, Location of Industries, pp. 37-38.

With these assumptions, Weber analyzed first the case where transport cost, in terms of resources used, determines location. Transport costs depend upon weight to be moved and distance. All other factors influencing total transportation costs, such as rate differences, were expressed in terms of these two by assuming, for example, the higher-rate line to be proportionately longer than a low-rate one.[1] This greatly simplified his work, but lessened the usefulness of his location theory in analyzing the impact of differential transport rates on regional industrial distribution.[2] Weber concluded that with transportation-orientation production will be drawn to points of minimum ton-mileage, given the locations of consumption and materials deposits. He distinguished ubiquitous materials, those found everywhere, from localized materials, those obtainable in a limited number of places due to economic and/or technological reasons. The use of ubiquitous materials would draw

---

[1]Ibid., pp. 41-42.

[2]One topic within political economy which location theory could examine is the impact of regulated freight rates on regional industrialization in the United States.

production towards the consumption point; use of localized materials would draw production to their deposits. A material could be distinguished further according to whether it entered the product without residue (a pure material) or with residue (gross material). Only pure materials would impart their full weight to the product.[1] The weight of materials used in production must be moved from their location to the production point; the product weight must be moved from the production site to the consumption point. Weber suggested the use of a locational figure to calculate the minimum transport-cost location, where each corner of the figure represents location of materials or consumption. Figure 1 illustrates a triangular locational figure with raw materials located at A_1 and A_2, and consumption occuring at A_3. By introducing weights of A_1 and A_2 per unit of weight of the product, the point of minimum transport cost could be determined geometrically.[2] P_o is the point where the three forces a_1, a_2, a_3 (representing weights of input per ton of

---

[1]Weber, Location of Industries, pp. 51-53.

[2]George Pick, "Mathematical Appendix", in Alfred Weber, Location of Industries, pp. 227-252.

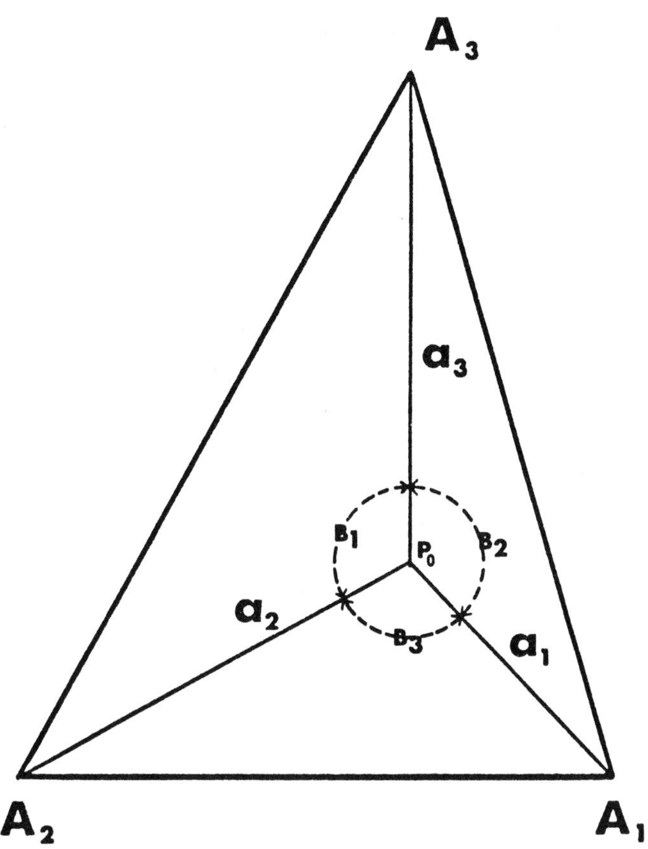

Figure 1

LOCATIONAL TRIANGLE

output and the ton of output) that are pulling from the locations of inputs and consumption, are in equilibrium. This point is calculated geometrically by drawing a weight triangle, G_1 G_2 G_3, whose sides represent the weight pulls of the triangle;s corners. The angles C_1, C_2, C_3 are supplements of angles B_1, B_2, B_3. P then must be on an arc from A_1 to A_2; it must also lie on an arc from A_2 to A_3, and on one from A_3 to A_1, so that $A_1 P_o A_2 = B_3$, $A_2 P_o A_3 = B_1$, and $A_3 P_o A_1 = B_2$. Construction of any two of these arcs provides point P_o.[1]

Weber defined the total weight to be moved in the locational triangle as the industry's "locational weight." He called the ratio of the weight of localized material to the weight of the product the "material index" and concluded that: (1) industries with a high locational weight generally would be attracted toward material sites and (2) industries with a material index of one or less would tend to locate at the consumption point.[2] Weber concluded that only the material

---

[1]Ibid.

[2]Weber, Location of Industries, pp. 60-61.

index and its composition, products of the temporary technical situation in the given industry, would determine location under transport orientation.[1] Weber's analysis, by assuming given demand and input locations, is certainly limited in the industries to which it is potentially applicable.[2]

After studying the static locational problem, Weber observed that with development, greater population, control over nature, and concentration could all be expected, with several consequences:

(1) With an ever-increasing demand for ubiquities (materials available everywhere) many of these would become localized materials (restricted in availability); this, by reducing the share of ubiquities in production,

---

[1]Ibid, p. 72.

[2]Weber noted that an increase or decrease in the rate level changes nothing at all in the whole picture--that the fundamental network of industrial orientation is independent of the general level of these transport costs; this is true only if level changes proportionally everywhere and also depends upon the nature of the demand curve; conceivably a change in transport costs will change quantity demanded which will influence the distribution of production if demand elasticities vary among consumption points. Isard and Lösch, among others, criticized this limitation of Weber's location theory.

would lessen the attraction of consumption points as production sites.

(2) Supplies of materials readily used in production would become exhausted, resulting in increased weight losses during production; this would strengthen the materials component in locational decisions.[1]

To facilitate examination of the individual firm's locational choice when geographical labor cost differences exist, Weber introduced the concept of isodapanes. He defined an isodapane as a curve around the minimum transport-cost location connecting all potential production sites having the same labor cost deviation from minimum transport cost. The critical isodapane would represent a deviation from minimum transport costs just equal to the labor cost economies involved in moving to a cheap labor site. If the actual lower-cost labor site were located on an isodapane closer to the minimum transport-cost site than the critical one, the economies in labor cost would exceed the increase in transport cost with location at the

---

[1]Weber, Location of Industries, pp. 73-75.

cheap labor source. In this situation, labor orientation would be substituted for transport orientation.[1] Weber noted that differentials in labor cost are not continuous; therefore to take advantage of lower labor costs, discontinuous jumps in location would be frequently necessary.

To analyze general conditions surrounding labor orientation, Weber introduced the labor coefficient, defined as the labor cost involved in producing one ton of locational weight to be moved. He observed that the higher the labor coefficient for an industry the greater would be the tendency for that industry to be labor oriented.[2] Here, as elsewhere, Weber did not take adequate note of the rising land rents which would result from greater concentration of industry in a given location, possibly offsetting economies offered by the cheap labor location. Weber indicated that movement to cheaper labor sources might result in substituting resource deposits closer to cheap labor for the sites originally hypothesized. Thus, divertability from the minimum transport-cost site would not be

---

[1]Ibid., p. 104.

[2]Ibid., pp. 111-113.

a simple function of the labor coefficient but might increase in greater proportion than that coefficient due to the possibility of substituting materials deposits.

Weber described three environmental conditions of labor orientation: (1) the distance between locational figures and labor locations; (2) the indices of economy of labor locations; (3) transportation rates. A greater population density, by reducing distances involved and increasing efficiency differentials among workers, would increase concentration at labor sites. Declines in transport rates, such as those wrought by the nineteenth century extension of railroads, would facilitate movement from handicraft centers. Mechanization, on the other hand, would reduce the labor coefficient and increase materials weight per ton of product, thereby favoring minimum transport-cost locations. Weber suggested that empirical study is necessary to determine which tendency has prevailed.[1]

Weber did not analyze extensively economies and diseconomies of concentrated location, though he did distinquish economies of scale for

---

[1]ibid., pp. 119-123.

the single firms from economies to the industry as a whole; Georg Pick, the author of the mathematical appendices to Theory of the Location of Industries did outline a mathematical approach to agglomeration economies.[1] Weber defined value added through manufacturing per total weight to be transported for a ton of output as the coefficient of manufacturing. He concluded that the tendency to agglomerate would vary directly with the coefficient of manufacturing, and that these tendencies would be inherent in the nature of the industry.[2] He did not outline a mechanism by which independent firms would be induced to agglomerate given the substantial moving costs that might be involved.[3]

Weber recognized the possibility of several stages of production being located separately from each other. This "splitting" was likely to occur, he noted, when raw materials with different sources were used at each stage of production.[4] With

---

[1]Georg Pick, Mathematical Appendix in Weber, Location of Industries, pp. 245-249.

[2]Weber, Location of Industries, pp. 162-67.

[3]See, for example, Isard's criticism in Isard, Location and Space (Cambridge, Mass.: The MIT Press, 1956), pp. 179-80.

[4]Weber, Location of Industries, p. 188.

increased mechanization in an industry, however, the technically differentiated tasks would tend to be reunited creating great concentrations of capital. As an example, Weber cited the iron industry, where the one-time separate steps of mining ore, producing steel, and rolling steel had been combined into one process.[1] He viewed this vertical reintegration due to increased mechanization as a revolution leading to a simpler locational orientation with ". . . units of locations of large-scale industries organized in combinations."[2]

After detailed examination of factors which could lead to deviation from his theoretically determined locations, Weber concluded that a clear understanding of location could be obtained by examining points of minimal transportation costs for individual production series, although labor and agglomeration economies must be considered for an understanding of the ways individual production processes are connected.[3]

---

[1]Ibid., p. 194.

[2]Ibid., pp. 195-96.

[3]Ibid., p. 210.

In his final chapter, Weber considered the location of manufacturing in an economic system. He relaxed the assumptions of given consumption, materials and labor sites, and infinitely elastic labor supplies at the given locations. He offered a historical-evolutionary approach, starting with a previously unoccupied country. Each stratum, beginning with agriculture, would be developed and supplied by each succeeding stratum.[1] The agricultural stratum, he explained, would be the geographical foundation for all others, fixing the places of consumption and materials deposits. Agriculture also would set the location of the primary industrial stratum which supplies agriculture's needs.[2] Weber suggested that there would then develop an industrial population to supply the needs of the primary industrial stratum; another layer of population would be involved in distributing goods, and still another population would exist which would only consume (officials, etc.); finally there would be an industrial stratum to

---

[1]Ibid., p. 214.

[2]Ibid., p. 215.

supply the latter two populations.[1] Although Weber did attempt a brief analysis of overall locational development, his emphasis was on the location of individual firms and industries; he concluded that in absence of labor deviations his pure rules would fix industrial locations. As for labor locations, Weber left that problem for other analysts.

Edgar Hoover also examined industry location rigorously. His study of the location of the shoe and leather industries began with an examination of general location factors.[2] Hoover first analyzed the location of extractive industries. Since these industries obviously would be tied to materials deposits locations, Hoover asked how the market area served by different deposits is determined. Hoover recognized that, due to economies and diseconomies of scale in extraction, the product price at any point would depend upon the entire market size. He called the line connecting the delivered prices at the market's edge for different

---

[1]Ibid., pp. 215-216.

[2]Edgar Hoover, Location Theory and the Shoe and Leather Industries, (Cambridge, Massachusetts: Harvard University Press, 1937).

market sizes the margin line. The consumer would purchase from the producer with lowest delivered price, assuming a homogenous product. Therefore, when each extractive site has a local market to itself, the market area boundaries would be determined by the intersection of margin lines. Hoover noted that the margin line and, hence, the market area, would be determined by the level and form of the transport rate schedule, the distribution of customers, demand elasticity for the product, and the nature of the cost function.

Hoover also utilized the margin line in analyzing the location of manufacturing industries, and combined it with Weber's isodapanes in the analysis. He defined isodapanes as loci of equal delivered price, where this delivered price would include the cost of moving materials to the production site and the product to the consumer. Although Hoover's approach was essentially integrative, he developed the individual components of the total location picture, including analyses of transport and labor orientation. He contended that as long as land is the only immobile production factor and there are no locational production cost differentials, location would become transport oriented.[1,2] In this analysis, Hoover followed the approach introduced by Launhardt and revived by Weber. Hoover criticized the latter, however, because Weber obtained the minimum cost location only for given materials and consumption sites without specifying why these sites were the appropriate ones to use.[3] Hoover concluded that the theory of market areas should be integrated with the orientation approach to achieve a complete theory of location. He acknowledged that orientation should come first in locational analysis, to determine whether transport costs are minimized at the market or at materials sites, and also to determine the size of transport costs relative to production cost differentials.[4] In examining orientation, Hoover prefered to use the term

---

[1]Ibid.

[2]Presumably Hoover also includes rising or falling marginal costs in suggesting location would be at the point of minimum transport costs.

[3]Hoover, Location Theory, p. 36. Weber had recognized this problem to some extent, and had discussed when alternative materials sites would be used.

[4]Within a theoretical framework, supply and demand factors should be considered simultaneously. Presumably Hoover suggests an initial examination of orientation to simplify actual locational choices.

transfer costs to that of transport costs. Transfer costs would include all costs which vary systematically with distance, such as transportation fees, insurance, product deterioration, etc.[1]

The typical pattern of manufacturing establishments would have one factory supplying a relatively large market area.[2] Utilizing the tools he has developed, Hoover analyzed and illustrated a variety of situations with different markets and materials locations. He determined theoretical market areas for various cost conditions, and similarly determined theoretical market areas when production is at materials deposits, at the market, and at intermediate low cost sites. He could then conclude where production would occur once cost conditions and the market are known. Production, of course, would occur at the site whose market

---

[1]Hoover credits the use of transfer costs to Bertil Ohlin who uses them in his Interregional and International Trade (pp. 142, 211). See Hoover, Location Theory, p. 39.

[2]However, in agriculture, numerous producers supply a given consumption point, generally an urban area. Agricultural location theory therefore is concerned with supply areas. A major contributor to agricultural location theory is Johann von Thunen in his Der isolierte Staat in Beziehung auf Landwirtschaft und Nationalökonomie, Hamburg, 1826.

area includes the given consumption location. In similar theoretical manner, he could determine the area supplied by various resource deposits, and could calculate which deposit would be used for given production sites.

Weber had discussed at length the possibility of production occurring neither at the materials nor at the consumption site. Hoover dismissed this possibility as unrealistic except where breakpoints occur in the transportation network.

Like Weber, Hoover considered the case of labor orientation, where labor cost saving is sufficient to pull location from the minimum transport cost site to a cheap labor site. At a cheap labor site, labor output per dollar of wages would be relatively high; hence, to find these sites, labor productivity as well as wages must be considered. Hoover considered two types of wage differentials, real and equalizing. The latter would reflect local differences in the cost of living, while the former would indicate labor immobility. Hoover appeared to have included skill differences in the equalizing category.[1,2] The major factor

---

[1]Hoover, op. cit., p. 70.

[2]Labor cost differentials due to unequal

in cost-of-living differences would be variation in food costs.[1] With a mobile labor force, then, labor-oriented firms, those with high labor inputs, would be attracted to agricultural surplus areas.[2] As labor-oriented firms crowd into cheap-food areas, the cost of living would be pushed up unless, presumably, agglomeration economies offset the effect of increased demand. As population in the area grows, market oriented industries would appear, further increasing living costs.[3] Hoover therefore concluded that activities in densely populated cities become increasingly market oriented unless there are economies of concentration for the

---

skills may be destabilizing. The higher wage area has the income for large educational expenditures, perhaps guaranteeing greater skills in the future.

[1]One might argue this point with Hoover when discussing affluent countries.

[2]Hoover analyzed budget materials; the prices of the budget materials, including costs of transfer determine the workman's cost of living. Hoover, Location Theory, p. 62.

[3]Again, whether or not living costs increase depends upon the net effect of agglomeration economies and diseconomies.

labor oriented firms.[1] Reductions in transport rates would tend to enhance labor orientation, according to Hoover. To the extent that labor cost is a function of food's delivered price, reductions in transportation rates should also decrease labor costs, however, which Hoover did not analyze.[2] Because transportation rates fell dramatically during much of the Industrial Revolution, Hoover suggested that the period witnessed an increase in the relative importance of labor costs, while more recently increased labor mobility has lessened somewhat labor's importance in locational decisions.[3]

Hoover illustrated the transfer cost advantages of a particular production site by a

---

[1]Information on basic versus nonbasic urban activities does suggest that as city size grows, an increasing portion of its production is for internal consumption.

[2]The question might then be one of comparing the decline in agriculture goods transport prices with the rates on other goods.

[3]With the introduction and widespread use of truck transportation, beginning in the 1920's, producers have had greater flexibility in deciding where to locate within a given area. Often the decision has been in favor of suburban rather than urban locations. This has created a difficult labor allocation problem, since many low income or unemployed individuals are located in the central cities.

transport gradient, relating distance from the production point to delivered price (sum of base price and transport costs). As Hoover explained, there is no such simple relationship when production cost differentials are examined; many factors would then determine production costs. Hoover felt it important to analyze transfer and production costs separately because of this difference. Weber had studied production economies from an orientation viewpoint. He had suggested that among the isodapanes (loci of equal total transfer cost), there was one at which transfer costs exceeded the minimum transfer costs by an amount equal to the economies in production costs offered at a nonminimum transfer cost location. Weber had called this the critical isodapane, and had noted that only if the particular location were within the critical isodapane would production economies outweigh transfer economies. In this case, production would occur at the alternative, nonminimum transfer cost location. Presumably both Weber and Hoover meant by these "production economies" something other than economies of scale, since both discussed the latter separately. Economies of scale would enlarge the market area for the lowest production plus transfer cost site, and only in that sense change locational decisions. Production cost differentials, as discussed by Hoover and Weber, were those associated with a particular location, for example due to weather conditions.[1]

In his study of production cost differentials, Hoover again criticized Weber for his emphasis on orientation; that is, for considering, via isodapanes, a situation in which market and materials deposits are given. Hoover argued that market area analyses should be used for this problem. He did so by illustrating the markets that would be served by production sites at the market, at materials deposits, and at intermediate low production cost locations under various cost conditions. Again, given the appropriate production cost and consumption site conditions, he could theoretically determine the best production location.

At this point Hoover mentioned the possibility of overlapping market areas. If homogenous

---

[1]This production cost difference might be considered to be a difference in resource costs. For example, sites with warmer weather conditions could be considered as having lower, or cheaper, fuel requirements.

products were produced by two companies, competition would occur only at the boundary of the market areas, where delivered prices of the two firms were equal. This analysis did not handle the case where all producers located in the same vicinity and shipped to a widely dispersed market.[1] Hoover added that market areas, in reality, do overlap due to imperfect product substitutability, freight cost absorption by sellers,[2] transport rate blanketing, and maintenance of a uniform retail product price over a large area. Sometimes, also, the insignificance of freight rates relative to total product price would permit overlapping market areas. With overlapping market areas, Hoover emphasized, market area analysis would be inappropriate.

Production costs vary with levels of output as well as with the location of production.

---

[1]The extent of scale economies should be important in determining how many and which firms share the market when all producers locate together.

[2]Freight cost absorption, if at all significant relative to the product price, should occur only when economies of scale in production occur, leading to marginal cost reductions greater than the marginal revenue declines. Alternatively, the freight absorber might be practicing a form of predatory price cutting with an emphasis on long run profits.

Hoover criticized Weber for lumping economies and diseconomies of concentration into one net agglomeration force, rather than carefully distinguishing the component forces. Hoover mentioned three types of economies due to central location. First, economies of scale, internal to the individual firm, would result in decreased average costs with increased quantity produced by one firm at a given site. Second, economies of localization, internal to the industry but external to the firm, would result in lowered per unit costs for all firms when industry output expands. Finally, economies of urbanization, which are external to individual firms and industries but internal for all economic activity within a given location, lower per unit costs when economic activity at a given location increases.[1] Since these urbanization economies would be external to the industry, they would appear to individual firms as factors creating production cost differentials in different locations. The first two types of economies, scale and localization, might be involved in studying location of a given industry. Hoover utilized the margin line

---

[1]Edgar Hoover, Theory of Location, pp. 90-91.

to determine the locational influence of economies of localization. With a number of firms located at point A, delivered price to the edge of the market would be a function of market area size. As market area expands, transport costs to the perimeter of the area would rise; as long as economies prevail, however, average costs of production would fall with increased output. When the latter fall more rapidly than transport costs rise, delivered price would fall as the market area grew, and the margin line would have a negative slope. Eventually, average production costs would fall less rapidly with increased output than transport costs rise or production costs will begin to rise, resulting in a positively sloped margin line. If the margin line from location A were to indicate a lower price in location B than the price at which output can be produced at B, location A would absorb the market at B. Hoover examined several factors which would reduce the slope of the margin line and thereby would increase the possibilities of localization. Falling unit costs as production at a central location rises would encourage central location; these economies would vary across industries and institutional arrangements of the particular industries and locations. Low transport rates would decrease the slopes of both the transport gradient and margin lines. Rates less than proportional to distance would create a flatter margin line further from the production center. A highly elastic demand curve for the product, suggesting large increases in quantity demanded if price falls with decreased marginal and average costs, would also flatten the margin line. The flatter margin line would make acceptance of the product in distant markets more likely.

Hoover also mentioned several potential diseconomies resulting from central location, including scarcity of particular inputs and greater labor organization creating higher wages; these would result in a rising margin line. Final product price is indicated by the price gradient, however, and not the margin line. The price gradient shows delivered price, for a given market area and base price, and hence must rise as distance from the production point increases.

Hoover did not completely distinguish economies of scale for one firm from the economies to the entire industry of central location. Although he did mention that the relationship between economies of scale and those of localization would determine how many firms locate optimally in a given place, he did not discuss the relationship of these centrally located firms to each other nor how the market area of each is determined. He did not really examine how, with firms separately located, some could be enticed to incur moving costs so that all might gain from localization economies.

Orientation would move the individual firm closer to materials or market sites to lessen transport costs, while economies of concentration would pull firms to a centralized location. Hence, Hoover foresaw possible conflict between the two. However, the two need not be in conflict; for example, many firms can locate near a materials source, e.g., the iron and steel firms in Pittsburgh. The localization advantages are related to economies of urbanization also, and these economies were discussed by Hoover more thoroughly in a later book on location theory.[1]

After applying location theory to the U. S. shoe and leather industries, Hoover discussed

---

[1]Edgar Hoover, The Location of Economic Activity, (New York: McGraw Hill Book Company, Inc., 1948).

needed research in the theory.[1] He suggested, for example, research into the impact of monopolistic competition on the individual entreprenuer's locational decisions. Since the forces of inertia would be likely to keep the actual short run pattern of location different from the long run equilibrium for long periods, Hoover urged the development of a short run location theory. In particular he advised an investigation of adjustment rates for various locational factors. He foresaw the use of public policy to improve the adjustment rates, especially when the adjustment was to temporary changes. Hoover acknowledged the difficulty of knowing when to use public policy in locational questions. He suggested its use whenever doing so would make the costs and benefits determining location more truly reflect social costs and benefits. That is, he wanted public policy that would internalize for private firms the externalities generated by their location decisions. He listed several cases where intervention might be necessary, including: 1) the case of wasteful exploitation of natural resources; 2) the case of monopoly

---

[1]Hoover, Theory of Location, pp. 294-99.

elements, for example under the basing point system, leading to nonoptimal results; 3) the case of urban concentration, in which Hoover felt private decisions are unlikely to be socially optimum; 4) and the case where real overhead costs of an industry include external capital, the costs of which the firm ignores in deciding potential locational changes.[1] Thus, Hoover concluded his comprehensive analysis and application of location theory with the acknowledgement that much remains to be done.

Several papers by American economists in the late 1920's and the 1930's were concerned with competition within a spatial context, and the impact of spatial differentiation on location. Harold Hotelling, in his 1929 article, "Stability in Competition," analyzed duopoly within a spatial context.[2] His duopolists were located on a line, each facing zero production costs and an inelastic demand, such that the quantity demanded is one

---

[1]Hoover, Theory of Location, pp. 281-300, passim.

[2]Harold Hotelling, "Stability in Competition," The Economic Journal XXXIX (1929), pp. 41-57.

unit in each unit of time for each unit length of the line. From this model, Hotelling concluded there will be a socially undesirable amount of clustering, with the firms moving as close as possible to their competitor and having the largest portion of unserved hinterland behind them. Given Hotelling's assumptions, this behavior would maximize profits for the mobile firm(s).

Arthur Smithies suggested in his article, "Optimum Location in Spatial Competition," that Hotelling's restrictive assumptions bias the decision in favor of centralized location.[1] Smithies especially criticized the inelastic demand assumption. Smithies examined the case of two competitors facing elastic demand curves, with each free to move his location instantaneously, and with constant marginal costs. Each competitor tries to maximize his instantaneous rate of profits with respect to total sales, so that each competitor's policy then depends upon how he calculates the rival will react to his price and location. Smithies hypothesized three extreme cases:

---

[1]Arthur Smithies, "Optimum Location in Spatial Competition," The Journal of Political Economy, XLIX, (1941), pp. 423-439.

1. each competitor assumes the rival will choose an equal price and symmetrical location to his own when he changes his price and location.
2. each competitor assumes the rival will choose the same price as his own, but will retain the current location when the original competitor changes price and location.
3. each competitor assumes the rival sets price and determines location independently of his action.[1]

In the first case, each producer would remain one-quarter of the total distance from the end of the line. In the second and third cases, Smithies concluded that the firms would locate closer to the center than in the first case, but not necessarily adjacent to each other. In particular, he suggested that the relation between freight costs and demand conditions would be crucial in determining equilibrium location and price. Smithies also examined the impact of small changes in marginal costs upon location and price.

---

[1]Smithies, "Optimum Location," pp. 489-90.

Neither Hotelling nor Smithies considered in detail a major factor limiting centralization of location; this is the resulting rise in land rents as competitors vie for locations in the "center" of the market. This rise in rent, combined with reduced demand from outlying consumers, would tend to pull producers away from locations near their rivals.

August Lösch, in his remarkable book, The Economics of Location,[1] followed part of Hoover's suggested inquiry, by examining locational decisions of the imperfectly competitive firm. However, Lösch indicated less hope for useful government intervention in locational decisions than did Hoover; and early in his book Lösch stated his desire to show that there is in general a rational natural order in the location of economic activity. Furthermore, this order would prevail most of the time without human intervention.[2] Lösch did not deny the usefulness of all intervention,

---

[1]August Lösch, The Economics of Location, translated by William H. Woglam with the assistance of Wolfgang Stolper, (New Haven and London: Yale University Press, 1954); it was first published in 1940.

[2]Lösch, Economics of Location, pp. 92-93.

but suggested that government understand the basic conditions underlying a locational equilibrium before undertaking policy. He encouraged intervention where it would move the economy more rapidly to its locational equilibrium.

A major contribution of Lösch to economic as well as location theory was the proof that, under specified conditions, a static locational general equilibrium would exist. This equilibrium would result from the interaction of individual economic units trying to maximize their advantage under the restraint created by society's attempts to maximize the number of independent economic units.[1] Lösch noted five conditions under which this behavior would lead to an equilibrium; he emphasized more the importance of understanding the conditions than of finding that a locational equilibrium exists. The conditions were that 1) the location of an individual must be as advantageous as possible; 2) the locations must be so numerous that the entire space is occupied;[2] 3)

---

[1]Lösch, Economics of Location, p. 94.

[2]Mills and Lav have questioned whether free entry will result necessarily in the space being filled. See Edwin Mills and Michael Lav, "A Model

in all activities open to everyone, all abnormal profits must disappear; 4) the areas of supply, production, and sales must be as small as possible; 5) at the boundaries of economic areas it must be a matter of indifference to which of two neighboring locations a given consumer (or producer) belongs.[1] Lösch explained that the equilibrium is born of locational interdependence, and ". . . can be understood only through a system of general equations of location."[2] The equations represented the five conditions listed above; careful counting indicated the feasibility of a general solution, since the number of unknowns equalled the number of solutions. What makes Lösch's proof especially appealing is that in his model he assumes imperfectly competitive firms.

Lösch's general equilibrium analysis has been criticized for its incompleteness, but his critics have given him credit for the scope of his work in introducing imperfect competition into the

---

of Market Areas with Free Entry," The Journal of Political Economy, LXXII (June, 1964), pp. 278-288.

[1]Lösch, Economics of Location, pp. 94-97, passim.

[2]Ibid., p. 98.

general equilibrium framework.[1] Lösch, for example, did not resolve the problem of individuals as consumers versus their role as producers. He did not explicitly introduce factor markets; he did not develop consumer demands from utility functions. In addition, Lösch merely counted equations and unknowns to conclude that a locational equilibrium exists.[2] Despite its limitations, his general equilibrium analysis, based on his theory of market areas for imperfect competitors (which is outlined below) is a monumental contribution.

Although he apparently found the mathematical abstractions of the general equilibrium intriguing, Lösch also discussed the more applied aspects of location theory in his extensive analysis of economic regions. He built his study of economic regions on the framework of market area analysis; his market area analysis is of major

---

[1]See, for example, Walter Isard, Location and Space Economy, (Cambridge, Mass., MIT Press, 1956), pp. 42-50.

[2]Another limitation, very difficult to overcome, is the static framework of the equilibrium analyses.

interest for analysis of firm and industry location.[1] Lösch assumed given cost curves in his market area analysis. All consumers, at the production site and in the surrounding hinterland, were assumed to have identical downward sloping demand curves for the product based on delivered price. The quantity demanded at a point thus depended upon price at the factory, distance from the factory to the given point, and transportation rates. Figure 2 illustrates the individual demand curve for a given product. If factory price were OF, quantity demanded at the factory site would be zero. At price OP, quantity demanded at the production site is OQ_1 or PQ_1. Quantity demanded by an individual falls as the individual's distance from the factory rises, due to rising delivered price. PQ_1 and the Q_1 and the Q_1F portion of the demand curve form a demand cone, for f.o.b. price OP, illustrated in figure 3. At that distance where delivered price is PF, given factory price OP, quantity demanded is zero; total quantity demanded at f.o.b. price OP is the area of the cone of revolution generated by the declining

---

[1]Op. cit., Lösch, pp. 105-108. What follows is a brief summary of Lösch's market area analyses.

curve of quantity demanded with increased delivered price (and hence increased distance) Q_1F, as shown in figure 3. Since the demand curve FS in figure 2 is only one individual's, the area generated by Q_1F must be multiplied by a population density factor to obtain total market demand at the f.o.b. price OP. This analysis assumes transportation costs and population density to be equal in all directions. Utilizing transportation rates, the distance from the factory of the point where quantity demanded is zero (the periphery of the market area) can be determined. Thus the market area can be calculated for factory price OP. Utilizing the same approach, total regional demand and market area could be calculated for all levels of factory price;[1] further, total and marginal revenue could be calculated for various factory prices. When combined with cost data, these revenue figures would yield the profit maximizing level of output, price, and market size for the firm. Given Lösch's assumptions for a general equilibrium we could conclude that the market area will be just large

---

[1]This assumes, of course, that firms located elsewhere do not alter their prices with changes in the original firms' price, or that there are no other firms.

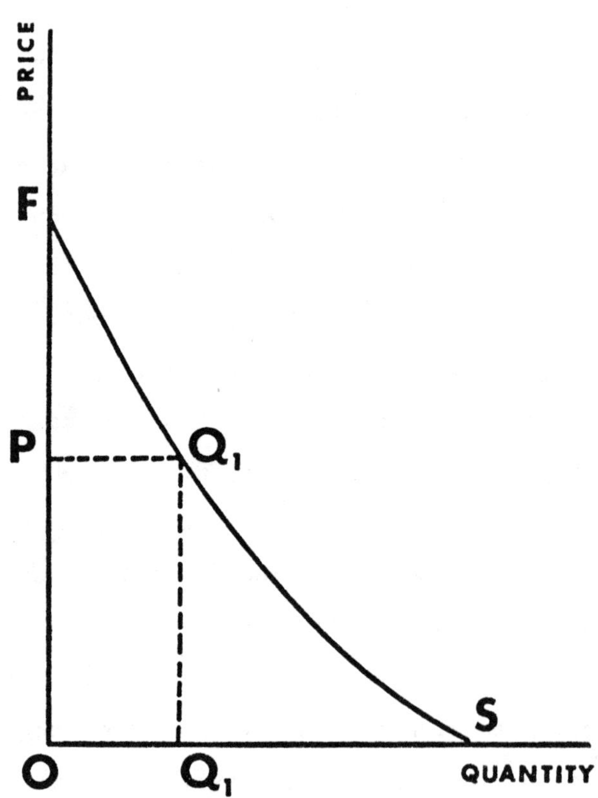

Figure 2

INDIVIDUAL DEMAND CURVE FROM

LÖSCH'S LOCATIONAL ANALYSIS

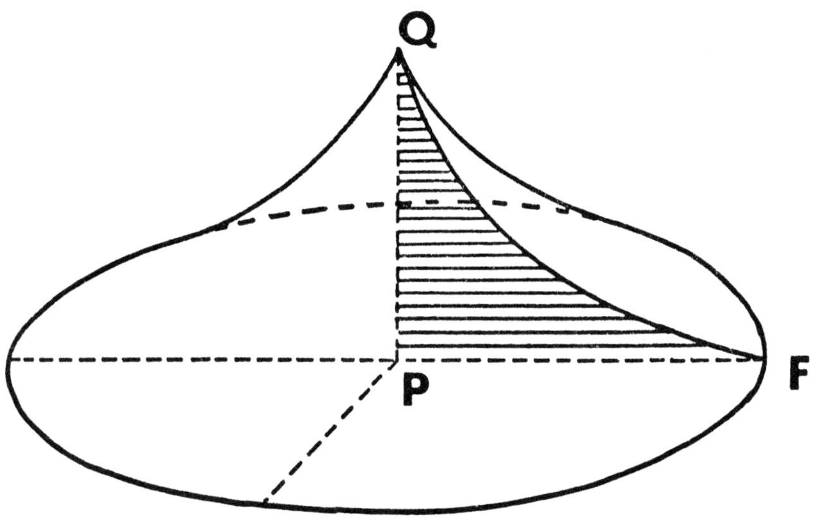

Figure 3

DEMAND CONE FROM LÖSCH'S

LOCATIONAL ANALYSIS

enough for total demand to exhaust economies of scale in production.[1]

In his analysis, Lösch explained that although a circular market would minimize transport costs, such a configuration would leave unserved gaps in the homogeneous plain he assumed to exist for his regional analysis. Hexagonal market areas would fill the entire market area with the lowest transport distances of any space filling geometrical figures; hence, Lösch concluded that the market areas would be hexagonal.[2]

From his examination of market areas for an individual product, Lösch determined that nets of hexagonal market areas would appear for every producer group. The metropolis would emerge as the common center of the nets of market areas.[3] Thus, Lösch concluded that even assuming no cost differences, across the physical setting a

---

[1]Of course, it is possible that the profit maximizing quantity is zero, suggesting that consumers must be self sufficient if they desire the product. See Lösch, p. 108.

[2]N firms can exist with unserved space left over, if the minimum sized efficient firm is too large to permit N+1 firms. See Mills and Lav, "Model of Market Areas," pp. 278-88.

[3]Lösch, Theory of Location, pp. 124-130, passim.

systematic arrangement of economic activity would emerge in the form of economic regions, distributed throughout the world in network fashion. He analyzed these regions and their relations to each other at length.[1,2]

Although Lösch's market area discussion helped in the analyses of firm and industry location, he did not spend much time on firm location. He noted early that his concern is more with the interdependence of locational decisions. He did, however, briefly discuss the impact of various spatial price policies on market area size, and concluded his spatial price policy commentary by noting the vital role of price in ordering

---

[1]Lösch apparently felt that metropolitan areas together with their subcenter and hinterland would be diversified and self-sufficient. His empirical findings showed more specialization than he apparently had expected. However, more recent studies seem to indicate that the larger a city, the less important are exports as a fraction of total production.

[2]In examining the metropolis more closely, Lösch suggested the importance of externalities. Lösch suggested that New York City, for example, has a tremendous advantage in its numbers, which explains a good part of its business (Lösch, p. 82). This seems somewhat to avoid a major locational question: why did New York rather than, say, Baltimore, acquire such numbers originally?

spatial relations.[1]

In his brief exposition of firm location, Lösch dismissed much earlier work for its emphasis on one-sided orientation, though he found this orientation unsurprising. For example, transport orientation could be expected because transport costs show great spatial regularity. Lösch did suggest that location at the point of minimum transport cost becomes meaningless once market area boundaries become flexible. While recognizing the value of Weber's work, he pointed out the unrealistic nature of Weber's assumption of completely inelastic demand curves. Lösch discarded transport cost, production cost, and gross receipts orientation, and substituted profit orientation for them.[2] Lösch did not really set up a model of firm and/or industry location; while his market area analysis touched on this problem, it over-simplified, quite deliberately, the cost side of the locational problem.[3]

---

[1]Lösch, <u>Theory of Location</u>, p. 189.

[2]Ibid., pp. 17-18.

[3]Again, Lösch's intent in market area analysis was not to study the individual firm location but rather to show how agglomerations of

Although Lösch emphasized location according to maximum profits, he did observe in a footnote that

> . . . Next to a drop in the level of freight rates to about 1/10, differential railway rates according to product and distance constituted the most significant change for the location pattern since the exit of the horse and wagon which, like the autotruck, charged freight essentially in proportion to weight and distance.[1]

In his empirical analysis, Lösch also recognized explicitly that locations of natural resources, to whose sources production is necessarily bound, do have special weight in the locational decision.[2] He summarized his approach to firm location by suggesting that location theory can be useful in determining the general region within which a firm should locate. However, he also cautioned that ". . . There is no scientific and unequivocal solution for the location of the individual firm, but only a practical one: the test

---

economic activity would occur even with homogeneous distribution of resources. With some economies of scale in production and positive transportation costs, concentration of production occurs. In a sense, then, his model emphasizes a type of transportation-orientation.

[1]Lösch, Theory of Location, pp. 170-71.

[2]Ibid., pp. 255-256.

of trial and error."[1]

Edwin Mills and Michael Lav questioned portions of Lösch's market area analysis in their article, "A Model of Market Areas with Free Entry."[2] They analyzed a market area model, assuming uniform space, and concluded free entry need not result in space-filling, hexagonal market areas. Lösch assumed that competition from free entrants into the industry would force the preferred circular market areas into hexagons so that the entire space would be filled. To disprove Lösch, Mills and Lav calculated profit algebraically with different market area sizes and factory prices for triangular, square, hexagonal, and circular markets. They plotted the relationship between market size and profits for these market shapes; they found the circular market shape to be the most profitable for any given market size. Mills and Lav then pointed out that production could be so unprofitable that only the most profitable market shape, the circle, would be utilized.

---

[1]Ibid., p. 29.

[2]Op. cit., Mills and Lav.

With circular market areas, some portion of the plain would not be served by central producers. They added also that whether free entry results in hexagonal or circular market areas would depend upon the values taken by the parameters of the profit model. The major parameters are the constant term and slope of the demand function, the transportation rate, the constant marginal cost, population density, and the level of fixed costs. Depending on the value of these parameters in their model (which assumes a linear demand function, constant marginal cost, and declining average costs) either spacefilling hexagonal or nonspace-filling circular market areas would result. They then generalized to suggest that the only possible market area shapes under their assumptions are circles and regular polygons with 6n sides where n can equal 1,2,3. . . . Utilizing their model, Mills and Lav also examined whether resources would be efficiently allocated between transportation and production costs, given free entry and profit maximization. They made the restrictive assumption that as firm size grows, f.o.b. price would fall sufficiently to keep average demand per family constant. They concluded that larger firm size and market area, resulting in lower average product costs due to economies of scale, would be preferable to the free entry result.[1]

Walter Isard, prominent in regional economics, attempted a general theory of location in his book, Location and Space-Economy.[2] He reviewed existing location theory extensively, suggesting four major directions in which location theory could develop. One possible approach was empirical, amassing new data and reexamining old material. A second method was to pursue the pure abstractions of Lösch. Still another possibility was to split the production process into two stages and examine especially closely the transportation stage, such as Koopmans and others had done.

---

[1]However, by restricting price to declines sufficient to keep average demand per family constant, with rising market area size, the authors guarantee little or no concern for transportation costs. The result, under economies of scale, has to prefer larger market area; the only counterweight to increased size is transportation cost. They also suggest that market size is limited by demand falling to zero on the market periphery as delivered price rises with market size. Even before this, the boundary could be set up by other firms reaching the consumer at lower delivered price.

[2]Walter Isard, Location and Space-Economy: A General Theory Relating to Industrial Location, Market Areas, Land Use, Trade, and Urban Structure, (Cambridge, Mass.: MIT Press, 1956).

Finally, Isard suggested his development of a general theory, which he hoped would encompass the major contributions of such theorists as von Thünen, Lösch, Weber, Predohl, and Ohlin.[1] Isard noted that monopolistic competition is inevitably present in spatial relations, and that a broad theory of imperfect competition would encompass location theory. Similarly, a general trade theory could be considered identical to a general theory of space and location, because of the often neglected importance of spatial relationships in trade.

Isard quickly suggested that if there is any rationale to location economics, it is due to certain regularities in the variations of costs and prices over space. These regularities are likely because of transportation costs. After investigating empirical evidence, Isard concluded that there are enough regularities with changes in distance to warrant theoretical consideration of spatial economics. He therefore continued development of his general theory, based on two major tenets. First is the substitution principle, used

---

[1]Ibid., pp. 24-54, passim.

by Predohl in locational analysis,[1] and familiar to general equilibrium theory. Under this principle, substitution among inputs would occur until the ratio of their marginal products would equal the ratio of their prices. Second is the transport input, defined by Isard as the movement of a unit weight over a unit distance. With these concepts, Isard hoped to fuse his location theory with existing production, price, and trade theory.

In the mathematical formulation of this theory, Isard described a general spatial transformation function.[2]

(1) $\Phi(Y_1, \ldots, Y_K, M_A S_A, \ldots, M_L S_L; X_{K+1}, \ldots, X_n) = 0$

where

$Y_1, \ldots, Y_k$: quantities of various inputs

$M_A S_A, \ldots, M_L S_L$: transport inputs

$M_A, \ldots, M_L$: weights of products, materials

$S_A, \ldots, S_L$: distances products, materials are moved

---

[1]Andreas Predohl, "Das Standortsproblem in der Wirtschaftstheorie," Weltwirtschaftliches Archiv, Vol. XXI, (1925), pp. 294-331, cited by Isard, Location and Space-Economy, p. 32.

[2]Isard, Location and Space-Economy, p. 222.

$X_{K+1}, . . ., X_n$: quantities of various products. Isard explicitly set transportation inputs apart in this function, since their change is fundamental in spatial economics. Lefeber has criticized Isard's treating transport inputs with a single set of symbols; more complete analyses would include supply and demand equations for transportation.[1]

Isard investigated the Weber minimum transport cost problem, Löschian market area analysis, and agricultural location theory, using his transformation function, substitution analysis, and notion of transport inputs.[2] He assumed a continuous space-economy, and ignored discontinuities in the transportation network, topographical variations, and economics of agglomeration. Isard recognized that these discontinuities might make substitution in the large more likely than substitution in the small. However, he also suggested that the degree of discontinuity not be exaggerated,

---

[1]Louis Lefeber, Allocation in Space: Production, Transport, and Industrial Location, (Amsterdam: North-Holland Publishing Co., 1958), p. 5.

[2]Isard, Location and Space-Economy, pp. 221-253 passim.

and pointed out the paucity of empirical studies examining the extent of spatial continuity and discontinuity.[1] From the application of his model, Isard discerned a basic, almost intuitively obvious principle, that

> . . . the marginal rate of substitution between any two transport inputs, however the transport inputs or groups of transport inputs may be defined, must equal the reciprocal of the ratio of their transport rates, social surplus (however defined) less transport costs on all other transport inputs being held constant.[2]

Isard proposed that this principle fuses existing separate partial location theories, and also allows location theory to be stated in a form similar to that of production theory.[3]

Martin Beckmann's Location Theory[4] contained a concise summary and mathematical

---

[1]Ibid., p. 251.

[2]Ibid., p. 252; it would seem that the principle could be extended to include marginal rates of substitution between transport and non-transport inputs; Isard seems to suggest essentially this on pp. 252-253.

[3]Ibid.; however, the generality of the assumptions surrounding the model lessens its usefulness for empirical work, as Isard recognizes.

[4]Martin Beckmann, Location Theory, (New York: Random House, 1968).

exposition of existing location theory, accompanied by Beckmann's extensions of that theory. The book covered a wide range of locational problems with a brevity requiring close scrutiny by the reader. The third chapter on location of an industry and the seventh chapter, "Locational Effects of Economic Growth," are most pertinent to a study of industry location.[1] Beckmann noted that although locational distribution of an industry is the classic location problem, there is no simple solution to the problem, depending, as it does, upon the situation in the given industry.[2] Beckmann distinguished four basic situations:

1. Producers and consumers concentrated at point locations and in point markets
   1a. Single point
   1b. Multiple points
2. Producers concentrated in points, consumers extended through an area (market area)
   2a. Single point
   2b. Multiple points

---

[1]Ibid., pp. 25-58, pp. 113-125, passim.

[2]Ibid., p. 25.

3. Consumers concentrated, producers dispersed (supply area)
   3a. Single point (central city)
   3b. Multiple points
4. Both producers and consumers distributed through an area: continuously extended market.[1]

The iron and steel industry seems most likely to fit into either category (1b) or category (2b). Which of these categories is appropriate for any given industry is a question to be explained by location theory, Beckmann noted. He also felt that both the time period involved and market structure influence locational decisions.

In examining location of an industry, Beckmann outlined possible spatial pricing policies of firms with differing market power, and the impact of these policies on market area size. For his analysis, he assumed a linear individual demand function, and constant average variable production

---

[1]Ibid., p. 26. Beckmann gives credit for much of this classification to L. Miksch, "Zur Theorie des raumlichen Gleichgewichts," Weltwirtschaftliches Archiv. Vol. 66 (1951), pp. 5-50.

costs.[1] He calculated for monopolistic firms the profit maximizing discriminatory price, uniform mill price, and uniform delivered price; only the uniform mill price is a truly nondiscriminatory spatial price. In all cases the boundary of the market area would be reached as soon as either quantity demanded falls to zero due to rising delivered price or price just equals marginal cost (including transportation costs) and the latter is rising faster than the former. The largest profit maximizing market area would occur under discriminatory pricing. In this case, the firm would vary price according to location, and, given Beckmann's assumptions, would maximize profit by absorbing one half the freight to non-adjacent customers. The two other pricing policies would both result in market areas three-quarters the size of the maximum market area size. The lowest profit maximizing mill price would occur under a uniform mill price policy; the highest mill price, under

---

[1]Ibid., pp. 30-32. Beckmann is not always clear in his use of marginal cost. In the figure (3.1, p. 32), he apparently included transport costs in c, marginal costs. Elsewhere (p. 30) he assumed average variable costs and hence marginal (production) costs, to be constant.

a uniform delivered price policy. Beckmann analyzed the same pricing policies for duopoly, with somewhat less definitive results, since the rival's reaction to a particular policy must be considered. He suggested collusion as an alternative to price rivalry under oligopoly, and noted that under mill pricing or discriminatory pricing the agreement might be either to set uniform mill prices or to set up a basing point system.[1] Under collusion, firms might also agree to split the entire region into separate market areas, each allocated to one firm; the firm then would have monopoly power within that market area. Under monopolistic competition, with negligible transport costs, Beckmann found the firm pricing by the same principles as in nonspatial circumstances. Assuming negligible transport costs would seem, however, to assume away much of the differentiation that distinguishes monopolistic competition. Finally, Beckmann examined perfect competition, a market form not generally deemed consistent with spatial analysis. Beckmann proposed that 1) if customers are indifferent to time costs

---

[1]For more information on the basing point system, see the final chapter.

in shopping, buying from the cheapest f.o.b. supplier, or 2) if transport costs are not sufficient to outweigh production cost advantages in one location resulting in concentration of many producers at that location, perfect competition might result. Then prices (presumably f.o.b.) would be equalized and brought down to the level of marginal costs of the least efficient surviving producer.

Beckmann also studied relocation of a firm and entry of new firms into existing markets under various market conditions and pricing policies, with his major concern being the resulting pattern of market areas. As a rule (though with exceptions), Beckmann found that free entry, yielding the densest packing of firms possible, would generate a long run equilibrium of hexagonal market areas.[1]

Beckmann assumed imperfect competition in much of his analysis, including that on relocation. However, he suggested that the firm maximizes profits by setting price equal to marginal total cost (marginal production plus marginal transport costs), rather than by equating marginal revenue and

---

[1]Beckmann, Location Theory, pp. 41-44, passim.

marginal cost. However, he essentially equated marginal revenue and marginal cost in some of his calculations; the switching is rather confusing, unless one assumes a horizontal demand curve at the prevailing market price.

In discussing market area size and efficiency in resource allocation under mill pricing, Beckmann concluded that market areas may be inefficiently large because the firms minimize production costs while ignoring transportation costs.[1] This probably overstates the case; if the consumer has to pay the transport costs, he will turn to the firm's competitors as its price plus transport costs rise above that of the rivals.

Although Beckmann's discussion of general equilibrium in space did not directly deal with the immediate problem of industry location, some observations are useful. He noted that long run equilibrium would necessitate that no activity could be profitably relocated; no land could be more profitably used; capital would be earning the same rate of return everywhere; and no mobile resource could move to a location where the extra

---

[1]Ibid., pp. 47-8.

benefits outweigh the moving cost. He concluded that the frictional moving costs, among other things, would result in an inefficient spatial allocation of resources.[1] Certainly these costs slow down relocation; if the rule, however, were reworded to suggest that in equilibrium the present discounted value of all benefits from moving could not outweigh the costs of moving, a more nearly optimum resource allocation could occur.

Economic growth has locational effects; Beckmann probed these in his final chapter. Regardless of the pricing system, for example, increased demand would increase both prices and market area, where possible. As a result competition would be intensified. In the short run, output would rise until capacity is reached; at that point, further increases in demand would only raise prices.[2] With economic growth, transport costs generally have declined; this has tended to increase actual or potential sales areas. If, as Beckmann suggested, the demand for transportation

---

[1]Ibid., pp. 105-6.

[2]Beckmann, pp. 114-5. This intensification of competition assumes no collusion on the part of the firms to split up the additional demand.

is price elastic, then reductions in transport cost (increase in supply) would result in small price declines and relatively large increases in the quantity of transportation demanded. This would continue until congestion and other diseconomies occur. Beckmann observed that with growing demand, the entry and location of new firms would depend upon the rate of demand increase. With limited demand growth (less than tripling) new firms would tend to locate near existing ones; with demand tripling, new firms would have incentives to locate at new locations in-between existing suppliers.[1] This assumes identical costs at the various locations; it is quite likely, however, that land costs will be higher in densely settled locations than in newer ones, thereby discouraging entry in the former. However, some input or sales costs may be lower in settled areas due to external economies (of agglomeration), enhancing the attractiveness of older locations.

Finally, after discussing the feasibility and nature of equilibrium in a growing economy, Beckmann examined the impact of growth and other

---

[1]Ibid., pp. 117-8.

changes on industries requiring localized inputs. He observed that technical change tends to increase the degree of fabrication, increasing the weight of the final output relative to that of inputs. Hence, he would expect production, other than the initial stages of reducing and refining the material, to move away from resource deposits over time.[1] Beckmann noted that economic growth will generate more spatial homogeneity by permitting deconcentration and a tendency toward equilibrium of price structures. He concluded his book with the observation that we are just beginning to understand the locational impact of growth, and indeed, "---unsolved and imperfectly understood problems abound throughout location theory."[2]

Although we have examined only a few of the important works in location theory here, some conclusions about the status of the theory vis-a-vis industry location can be summarized. Weber's analysis of firm and industrial location is a major contribution emphasizing supply conditions.

---

[1]Ibid., p. 124. This conclusion is opposite to that of Weber, who anticipated movement toward resources over time.

[2]Ibid., p. 125.

Weber dealt primarily with cost (especially transport) minimizing locations, under fairly restrictive conditions and his analysis of the dynamics of locational development emphasized an historical-evolutionary approach.

Hoover corrected and extended much of Weber's analysis, and lessened the restrictiveness of Weber's assumptions. He also considered the locational impact of such supply conditions as economies of scale. In addition, Hoover extended greatly the graphical analysis of locational problems. He infused more reality into location theory through his analysis of market areas. However, it was Lösch in his limited analysis of firm and industry location who thoroughly examined market areas, especially their size and shape. He utilized the market area analysis primarily as a building block for his examination of economic regions, but his insights are useful in studying industry location. Lösch noted that the firm would be oriented by profits in its location decision rather than by any of the individual components of profit. He also added much in his discussion of spatial pricing policies and their impact on market areas.

Isard emphasized the role of transportation in his locational analysis by using a spatial transformation function which separated transport inputs from other inputs. Isard added to this Predohl's substitution concept to determine principles for optimal location.

Beckmann synthesized much of previous location theory into a more cohesive body, and also improved the resulting theory through careful mathematical analysis of the problems. His work is not as novel perhaps as earlier location theorists, and was not meant to be; however, his careful methods sometimes yielded conclusions different from those of earlier economists and do offer useful insights. His book is a helpful summary and extension of location theory.

Location theory thus provides an outline of the necessary framework for examining industrial location. However, most of the preceding authors did not set up models of industry location that were amenable to empirical solution. Therefore, it is necessary to incorporate the locational variables in a tractable model. Before handling this task, the industry under study is described.

CHAPTER 3: DESCRIPTION OF THE STEEL INDUSTRY IN THE UNITED STATES, 1879-1919

I. INTRODUCTION

Before specifying a model of industrial location for the American steel industry, some background on the technology and raw materials used in the industry is essential. This chapter first describes the technology of raw and semi-finished steel production, then discusses the necessary raw materials and their locations. The chapter then deals with another important locational influence, the pattern of steel demand. Direct observations of the location of steel demand for this period are unavailable; therefore, specific demand figures must be constructed as described in section 5 of the chapter. Other important elements in the steel industry's description include its susceptibility to cyclical fluctuation, the pattern of industrial concentration, and the actual location of the industry. These elements are covered in the final sections of the chapter.

II. TECHNOLOGY

Between 1880 and 1920 the iron and steel industry was transformed by the rapid rise in steel production and the relative demise in iron output. Table 3.1 indicates the shift from iron to steel in the late nineteenth and early twentieth centuries. The transformation of the iron and steel industry resulted from technological advances in the application of heat to raw materials, especially iron ore, which permitted steel to be produced much more cheaply than previously. An understanding of how new technology influenced the industry requires some information on the basic processes involved in iron and steel production.[1]

---

[1]Information on the technology of iron and steel production is taken from several sources, including Peter Temin, *Iron and Steel in Nineteenth Century America: An Economic Inquiry*, (Cambridge, Massachusetts: The M.I.T. Press, 1964); T. K. Derry and Trevor I. Williams, *A Short History of Technology*, (London: Oxford University Press, 1960), Frank Popplewell, *Some Modern Conditions and Recent Developments in Iron and Steel Production in America* (Manchester: University Press, 1906); James M. Swank, *History of the Manufacture of Iron in All Ages*, (Philadelphia: The American Iron and Steel Association, 1892); E. B. Alderfer and H. E. Michl, *Economics of American Industry*, (New York: McGraw-Hill Book Company, Inc., 1942); Douglas A. Fisher, *The Epic of Steel*, (New York: Harper & Row, Publishers, 1963).

Table 3.1

PRODUCTION OF IRON AND STEEL, 1879-1919

Year	Total Iron & Steel Products (Tons)	Iron Production (Tons)	Steel Production (Tons)	Percentage of Total Represented by Iron	Percentage of Total Represented by Steel
1879	3,411,562	2,353,248	1,058,134	68.98	31.02
1889	8,204,930	3,225,140	4,979,790	39.31	60.69
1899	14,538,231	4,030,387	10,507,844	27.72	72.28
1909	26,723,274	3,749,310	22,973,964	14.03	85.97
1919	36,211,947	2,184,968	34,026,979	6.04	93.96

This table is derived from the *Census of Manufactures* for 1880, 1890, 1899, 1910, and 1920. Because iron and steel output is not given separately for several of the years, the figures involve some manipulation. For example, the Census of 1900, 1910 and 1920 includes rolled iron and rolled steel in one category. To obtain rolled iron tonnage for those years, the amount of pig iron used in steel production and cast iron production is subtracted from total pig iron output. This difference is assumed to have been used for rolled iron production.

Iron is usually mixed with impurities in its natural state, iron ore. The metallurgy of iron and steel consists essentially in removing these impurities. Also, limited amounts of some impurities improve the quality of steel for particular uses, and hence controlled quantities are readded before the product is complete. Intimately involved in this process of purifying and/or strengthening natural iron is the use of heat in melting iron ores and oxidizing impurities. Many of the technological advances in the industry involved improvements in heat usage.

Nineteenth century iron and steel was produced in two stages. In the first, pig iron was produced from iron ore by smelting the ore in blast furnaces to remove oxygen. It was also possible to cast the molten product from this stage, either immediately, or after remelting, to produce cast iron. Cast iron, containing 2% or more carbon, was easily fusible, but could not be hammered or rolled into shapes. Molten pig iron was cast when shapes too intricate to be rolled or hammered were desired. Alternatively, the pig iron was used as an intermediate product, and in the second step, further refining removed carbon from the pig iron to yield wrought iron, usually by applying heat and stirring (puddling) the molten iron. Occasionally in the early nineteenth century, but with decreasing frequency over the century, wrought iron was made directly from the iron ore at bloomaries or bloomary forges. Refinery forges used pig iron to produce wrought iron, which was an almost completely pure form of iron (less than .3% of carbon), though it contained some slag. Wrought iron was a soft, malleable material. Initially it was hammered manually into shape for final use; later advances permitted machine rolling the wrought iron into desired shapes.

Very little steel was produced in the U. S. in the antebellum period. Steel is essentially pure iron with limited amounts of carbon added. The early products, blister steel and crucible steel, were very expensive and used only where a high quality product was essential. Blister steel was made by remelting wrought iron to add small amounts of carbon. Because wrought iron contained slag, it was not homogeneous in composition. The same non-homogeneity was shared by blister steel. Crucible steel was made, very expensively, by melting blister steel in small clay crucibles, each crucible holding about 60 pounds. Special clay, not readily available, was required for the crucibles. Antebellum technology did not allow any way to contain melted steel other than through the use of the small crucibles; the technology did not permit maintenance of the high temperatures necessary to keep the metal molten in large scale converters until the conversion from iron to steel was complete. In the 1850's Kelley, working in the United States, and Bessemer, in Great Britain, independently determined that a blast of air directed into molten pig iron would cause oxidation of the iron's impurities. As carbon is removed from iron, the iron's melting point rises. Earlier, in the production of wrought iron, as part of the iron solidified, the wrought iron developed the nonhomogeneous character typical of that product. Due to high temperatures created in the Bessemer process, the iron remained molten, and a uniform product resulted. By stopping the conversion before all the carbon was oxidized, steel, which is essentially iron with up to 2% carbon, was produced. Timing of the stop was difficult and a more uniformly high quality steel resulted when Robert Mushet discovered that all the carbon could be removed and then spiegeleisen added to the molten iron as a recarburizing agent.

This new airblast method of steel production, known as the Bessemer process, utilized no external fuel source once the pig iron was molten; rather, it depended upon the burning of the impurities in the iron to maintain high temperatures. The Bessemer process permitted large scale production of steel with virtually no fuel costs as long as molten pig iron was charged into the converter, but the process did require capital equipment substantially different from the equipment used to produce wrought iron.

The Bessemer process was utilized in the United States as early as 1864, though extensive use was delayed until the late 1860's. The demand for steel rails encouraged rapid growth of the Bessemer steel industry. There were, however, difficulties in the process that led to its demise vis-a-vis the open hearth process in the 1880's. The greatest difficulty was that the original Bessemer process used an acid lining in the converter; although some impurities were eliminated in the operation of an acid converter, phosphorus was not and phosphorus was detrimental to steel quality.

This meant that only very low phosphorous pig irons (no more than 0.1% phosphorus) could be used. Most of the U. S. iron ore supplies produced an iron with more than 0.1% but less than 1.5% phosphorus. Alternatively, a Bessemer converter with a basic lining could be used. This basic Bessemer converter required a high phosphorous pig iron since the basic converter generated heat by burning the phosphorus in the molten iron rather than the carbon. Thus, in the basic Bessemer process, pig iron produced from ores containing more than 1.5% phosphorus was needed. Since most American ores contained between 0.1% and 1.5% phosphorus, neither Bessemer process was amenable to using the iron ores most readily available in the United States. A final difficulty with Bessemer steel was its tendency toward fractures. These fractures resulted from nitrogen in the steel, which in turn resulted when the molten product was in close contact with air. Since the Bessemer process depended upon such contact, this defect was observed; however, its chemical source was not known in the late nineteenth century.[1]

---

[1]Temin, Iron and Steel, pp. 151-152.

The second major technological advance in the postbellum steel industry was the open-hearth method of steel production, introduced as early as 1866. It was an extension of the puddling process, via which wrought iron had been produced. Puddling was the manual manipulation of molten metal to keep its temperature high. The open hearth furnace involved higher temperatures than puddling and, because of the higher furnace temperatures, eliminated manual manipulation of the molten iron by the puddler. These higher furnace temperatures were achieved with regenerative stoves, which had been introduced by William Siemens. In the open-hearth furnace, the iron was exposed to flames from the burning fuel, usually a coal derived gas. The fuel was burned in regenerative stoves containing two compartments. While fuel was burned in one compartment, exhaust from the furnace was preheating the other. The heat from the burning fuel was added to the existing heat to create much higher temperatures than previously had been obtainable. The initial open-hearth furnaces were lined with an acid material and shared with the acid Bessemer process an inability to use high phosphorus pig iron. In

1875, Thomas and Gilchrist discovered that a basic lining in the blast furnace would combine with the phosphorus in the iron and that impurity would be carried off in the slag; they reported this discovery in 1879. The basic lining could be used in either Bessemer or open-hearth furnaces. Basic Bessemer furnaces generated heat by burning the phosphorus in the molten iron, rather than carbon, which meant that only iron ores containing more than 1.5% phosphorus could be used. The basic open-hearth method did not depend upon burning impurities as a source of heat, and hence could use any iron. Because most iron ores in the United States contained between 0.1% and 1.5% phosphorus, they were not suitable for acid open-hearth, acid Bessemer, or basic Bessemer steel production; however, basic open-hearth was easily produced by the available ores.

Temin, in examining the relative costs and characteristics of acid and basic Bessemer steel and acid and basic open-hearth steel, finds the basic open-hearth process both able to use a much wider range of American iron ores than the other three processes, and also producing a higher

Table 3.2

STEEL PRODUCTION, BY KINDS-TONS OF 2,240 LBS.
Percentage of Total of a Particular Kind is Given in Parenthesis Under the Tonnage
(Sum of Ingots and Castings)

Year	Bessemer	Open-Hearth	Crucible and Miscellaneous	Electric	Total
1879	889,896 (84.08%)	93,143 (8.80%)	75,275 (7.11%)	-	1,058,314
1889	4,385,365 (86.84%)	590,198 (11.68%)	74,130 (1.46%)	-	5,049,693
1899	7,528,267 (71.64%)	2,878,827 (27.39%)	100,750 (.95%)	-	10,507,844
1904	7,768,915 (56.82%)	5,820,397 (42.57%)	80,059 (.58%)	1,221 (-)	13,670,592
1909	9,180,133 (39.02%)	14,228,377 (60.48%)	100,263 (.42%)	14,426 (.06%)	23,523,199
1914	6,219,304 (26.57%)	17,081,375 (72.98%)	81,685 (.34%)	21,593 (.09%)	23,403,957
1919	6,946,939 (20.41%)	26,726,036 (78.54%)	64,245 (.18%)	289,759 (.85%)	34,026,979

Derived from Census of Manufactures, 1890, 1900, 1905, 1910, 1914, 1920.

quality product than the Bessemer processes.[1] These two factors, he feels, account for the replacement of Bessemer steel by open-hearth steel in the early twentieth century. This shift from Bessemer to open-hearth steel production, which was superimposed upon the late nineteenth century change from iron to steel, is indicated in Table 3.2.

In the first twenty years of the twentieth century, technological advance in the industry involved mainly modifications of processes developed in the late nineteenth century, although one additional mode of steel production, the electric method, was introduced. Siemens, as early as 1880, had produced steel experimentally in a furnace using electricity to generate heat. The first commercial electric steel plant had been built in Italy in 1896 and numerous other European plants, mainly dependent upon hydroelectric power, were in operation by 1900. For European steel producers the hydroelectric power, when available, was cheaper than mineral fuel. American electricity was usually steam generated, so fuel costs for electricity were not

---

[1]Temin, Iron and Steel, pp. 142-152.

lower than in conventional steel production in many American locations. The first American electric steel plant was in production by 1908; this process did have the advantage of being cleaner than those techniques which used fuel directly. Further, the electric technique permitted higher temperatures and greater heat control than either the Bessemer or open-hearth processes in the early twentieth century and hence allowed better mixing of steel alloys than the other two techniques. After World War I, the electric process was adopted increasingly for steel alloy manufacture.[1]

[1]Alderer and Michl, The Economics of American Industry, p. 48.

III. PROCESSING OF STEEL INTO SHAPES

Steel ingots were not generally the finished product produced and sold by the steel industry in the late nineteenth century. Just as wrought iron was typically rolled into some semifinished shape in rolling mills prior to sale, so was steel usually processed into semifinished shapes before being sold. Some molten steel was poured directly into intricate molds, producing cast steel when the molten steel cooled and solidified. More often, the molten metal was poured into more or less rectangular molds and allowed to solidify into ingots. The ingots, when evenly cooled to a desirable shaping temperature, were then further processed into semifinished shapes. Alderer and Michl list the four principal rough steel shaping operations utilized by 1900 as rolling, forging, pressing, and casting, with rolling the most important of these.[1] Steel ingots had to be hot before being rolled; fuel was conserved if the ingots were allowed to cool only to the desired rolling temperature before rolling rather than reheating cold ingots. The

---

[1]Alderer and Michl, The Economics of American Industry, p. 52.

fuel economies from controlled cooling followed immediately by rolling were obtainable only if the rolling mills were adjacent to the steelworks; this provided an advantage to the vertically integrated steel works and rolling mills. Among the steel shapes rolled in the late 1800's were bars, rods, bands, hoops, rails and plates. By the early twentieth century, long, thin wide sheets of steel were also being rolled.

IV. RAW MATERIALS

Production of pig iron requires two major raw materials inputs, iron ore and fuel, and a fluxing material, such as limestone. Steel production involves further processing of pig iron as outlined above, and therefore has essentially the same materials requirements as pig iron, though quantities of ore and coal per ton of steel will differ from those per ton of pig iron. Ideally, the production of pig iron and steel would occur at the simultaneous location of steel markets, ore mines and coal deposits. Such coincidence of the two major raw materials' locations and the market did not generally occur in the United States, as James Swank explains in the _Census of Manufactures_ of 1879:

> The average distance over which all the domestic iron ore which is consumed in the blast furnaces of the United States is transported is not less than 400 miles and the average distance over which the fuel which is used to smelt it is transported is not less than 200 miles . . . . The manufactured products themselves must be frequently transported long distances to find consumers.[1]

---

[1]James Swank, "The Manufacture of Iron and Steel," _Report on the Manufactures of the United_

By the late nineteenth century, coal and/or coke fueled most iron and steel production in the United States; thus by the postbellum period, although some charcoal iron was produced, the conversion to mineral fuel was nearly complete, especially for the large integrated iron and steelworks. Coal sufficiently pure to produce a high quality pig iron and steel was not ubiquitous. Additionally, the late nineteenth century practice of hard driving in the blast furnace to increase output of pig iron required a fuel that would burn rapidly, but not be crushed. Good coke is such a fuel; the best available sources of good coking coal in the late nineteenth century were the Connellsville coal fields, less than forty miles southeast of Pittsburgh. These fields provided sulfur free coal for local coking ovens and were the major source of fuel for the turn of the century American steel industry.

Southwest of the Connellsville coal field was the Pocahontas coal field of West Virginia. Not nearly as large as the Connellsville field, the

---

<u>States at the Tenth Census</u>: (Washington, D. C.: Government Printing Office: 1883), p. 142.

Pocahontas field was the third largest source of coal for the United States steel industry in the early twentieth century.[1]

Coal mines in central Alabama provided fuel for that state's iron and steel industry, centered around Birmingham. Although abundant coal supplies were available in Alabama, the Alabama coal was not high enough in quality to be shipped very far from the mine.

In both the eastern and western parts of the country, local coal sources were used by local iron and steelworks. East of the Allegheny Mountains in Maryland, Pennsylvania, and New York, some local coal sources were used to produce iron and steel. The major producer of iron and steel located west of the Mississippi River before 1900, the Colorado Fuel and Iron Company, used Colorado coal to produce iron and steel in Pueblo, Colorado, about a hundred miles south of Denver. Neither the east coast nor the Colorado coal was used in iron and steel production except by local furnaces and

---

[1]United States Commissioner of Corporations, Report of the Commissioner of Corporations on the Steel Industry; Part III: Cost of Production, (Washington: Government Printing Office, 1913), pp. 68-70.

steelworks.

Most late nineteenth century American coke was produced in beehive ovens. By the turn of the century by-product coke ovens, which recovered the chemical impurities in the coal as a saleable product and could produce relatively pure fuel from impure coal, were available for coke production. This alternative to beehive coke production somewhat lessened the steel industry's dependence on the pure Connellsville coal supplies, although Connellsville was still a major fuel source in 1919.

Just as coal supplies were localized in the late nineteenth century, so also were sources of high iron content iron ores. Iron ore was sufficiently abundant in North America to allow dispersed small scale production of iron in the early nineteenth century. Sufficient quantities of iron ore to support large scale steel production were not readily available. Throughout the period 1879 to 1919, the major domestic sources of ore were the Great Lakes ore mines. After the Civil War, the Lakes ores, including deposits in Michigan, Wisconsin, and Minnesota, were increasingly substituted for local ores, both around Pittsburgh and in the East. The Great Lakes ores had been discovered before the Civil War, but exploitation of these ore fields was slowed by their isolated location. Improvements in transportation, which included canals, railroad lines, and mechanization in loading and unloading ore trains and boats, permitted greater use of these ore deposits. In the 1870's, the Lake ores were one-quarter of the iron ore mined in the U. S.; by 1890, they constituted one-half the total, and by 1900, that figure had risen to two-thirds.[1] To get the ore from the field to Pittsburgh required train transport from the field to Lake Superior, transshipment to an ore boat, a trip down the Lake, transshipment to a train at Erie, Pa., and then shipment by train to Pittsburgh.

By the end of the nineteenth century, the Lakes ores, most of which were sufficiently free of phosphorus to be used in acid Bessemer steel production, were used widely both east and west of the Allegheny Mountains. Their only substantial competition in the North came from foreign ores; these were used primarily at east coast steelworks. Ores were imported from Spain and Algiers by 1880

---

[1]Temin, Iron and Steel, p. 195.

and the first ore shipment arrived from Cuba in 1894. The Sparrows Point steelworks, located on the Baltimore harbor, was supposedly built to take advantage of low water transportation rates on foreign ores.

Central Alabama was the only iron producing area in the United States at this time which had both extensive coal and ore supplies locally available. Alabama iron ores were high in phosphorus and therefore generally inadequate for Bessemer steel production. However, this restriction was not applicable for basic open-hearth production. These ores were lower in iron content than Great Lakes ores; they were not used beyond the local iron and steel industry of the Birmingham area.

For industries which process bulky raw materials into semifinished products, the weights of the different raw materials per unit of the product are important in locational decisions. Walter Isard, after examining coal and ore requirements per ton of pig iron in the nineteenth and twentieth centuries concludes that pig iron production is becoming less drawn to fuel and ore sources and more sensitive to demand sites. He predicts this movement away from raw materials will occur because of falling input requirements per ton of output.[1] Isard's data concerns primarily coal requirements per ton of pig iron; he gives ore inputs per ton of pig iron for one or two years.

Using census data and Isard's information, the coal input per ton of steelworks' output, α_c, and the ore requirement per ton of steelworks' output, α_r, were calculated for each major census year, 1879-1919. Values for α_c and α_r are listed in Table 3.3. Evident in the table is the overall decline of raw materials weights per ton of output. Given the rough average requirements shown in Table 3.3, ore tonnage per ton of output fell 77% over the forty year period, while coal requirements fell 67%.

Open-hearth steel converters could use large quantities of scrap steel as input. As the open-hearth process became increasingly dominant, scrap steel became a more important raw material. However, as of 1919, a large portion of scrap used

---

[1]Walter Isard, "Some Locational Factors in the Iron and Steel Industry Since the Early Nineteenth Century," Journal of Political Economy, LVI (June, 1948), pp. 203-217; Walter Isard and William Capron, "The Future Locational Pattern of Iron and Steel Production in the United States," Journal of Political Economy, LVII, (April, 1949), pp. 118-133.

by steel producers was generated by the same producers and the locational impact of scrap steel was still minimal.

The final component of iron and steel production, the fluxing agent, is nearly ubiquitous. Therefore, its location has been fairly unimportant in the geography of iron and steel industry.

Table 3.3

AVERAGE TONNAGE OF ORE (α_r) AND COAL (α_c) PER TON OF STEELWORKS' PRODUCT, 1879-1919

Year	Coal Per Ton of Steelworks' Output α_c	Ore Per Ton of Steelworks' Output α_r
1879	3.93	2.11
1889	2.77	1.59
1899	2.51	1.79
1909	2.55	2.97
1919	1.31	1.16

Calculated from figures given in the *Census of Manufactures* for 1880, 1890, 1900, 1910, 1920 and from estimates by Walter Isard, "Some Locational Factors. . . ." The ore requirement for 1909 is an unexpected deviation from the general pattern, but several calculations yielded the same result. About this time, steel producers were beginning to use benefication techniques to obtain more metal from lesser quality ores. Perhaps the increase in α_r encouraged the development of benefication, which in turn reduced α_r by 1919.

V. DEMAND

A. Products of Steelworks and Rolling Mills

In 1879 rails dominated the output of finished and semifinished steel products, with virtually all rails produced from Bessemer steel. By the end of the nineteenth century, alternative uses of steel were challenging the dominance of rails. Although some steel was cast throughout the era being considered, most raw steel was formed into ingots which were then rolled into rails, bars, rods, structural shapes, plates, and sheets. Temin defines eight major product categories for the nineteenth century iron and steel industry; they are (1) rails; (2) bars and rods: products of uniform cross section, excluding those used for wire production; (3) plates and sheets: flat products obtained by rolling, excluding those used for making nails; (4) wire rods: rods used to make wire by drawing; (5) skelp, etc.: strips used for making welded pipe, with hoops, bands, cotton ties often included because of their similar shape; (6) structural shapes: bars of various cross sections, shapes and sizes, though smaller sizes are included with bars and rods: (7)

nail plate: plates from which nails were cut; (8) other products: a residual category.[1] Table 3.4 gives the relative output of these categories, with skelp and structural steel combined, and also all plates and sheets combined. As can be seen from the table, rails declined in importance as did bars and rods. Wire rods, used to produce drawn wire from which barbed wire and wire nails were made, did increase in importance with the western agricultural demand for barbed wire and the replacement of wire nails for nails cut from iron and steel plates. The prominent position of rails was not replaced by any single use of steel. Rather, several semifinished products, demanded for urban construction and manufacturing, shared in later years the role played by rails in 1879. The declining importance of rails was foreseen by Andrew Carnegie, head of the largest steel producing firm in the United States at that time, when he commented in 1885, "The railroad system is practically developed . . . . The rail mills must

---

[1] Peter Temin, "The Composition of Iron and Steel Products, 1869-1909," Journal of Economic History, 23 (December, 1963), pp. 449-450.

Table 3.4

PRODUCTS OF THE AMERICAN STEEL INDUSTRY, 1879-1919. (TONS OF 2240 LBS.)

Year	Tonnage of Rails	Tonnage of Bars & Rods	Tons of Wire Rods	Tons of Structural Steel & Skelp	Tons of Plates & Sheets
1879	1,217,497	979,041	-	225,768	293,450
1889	2,091,686	1,761,029	-	788,987	731,116
1899	2,251,337	2,493,159	917,000	2,052,172	1,995,046
1909	2,844,061	3,975,606	2,295,000	4,207,916	3,991,013
1919	2,184,747	5,180,425	2,484,428	5,115,525	7,905,496

now adapt themselves to other purposes."[1] The Carnegie Company's new steel facilities were geared to structural steel production by the mid 1880's. Structural steel was used for bridge construction but, more importantly, also for construction of urban multistory buildings. Kirkland, in examining the economic impact of growing U. S. urbanization in the late nineteenth century, suggests that growth of American cities regenerated the steel industry, substituting the urban demand for structural steel for the lagging transportation demand for rails.[2] With the construction of the ten story Home Insurance Company building in 1885 and the twenty story Masonic Temple in 1891, both in Chicago, the feasibility of tall metal framed buildings was proved, and structural steel production increased.

Large urban areas began to invest in municipal water and sewage systems in the 1880's. These systems required considerable amounts of

---

[1] Joseph Frazier Wall, _Andrew Carnegie_, (New York: Oxford University Press, 1970), p. 654.

[2] Edward Kirkland, "Building American Cities," in Thomas C. Cochran and Thomas B. Brewer, ed., _Views of American Economic Growth: The Industrial Era_, (New York: McGraw Hill Book Company, 1966), pp. 15-20.

steel pipe. Wrought iron pipe had been used in the late 1870's and early 1880's to carry petroleum. Tubes were in demand in the 1880's for carrying natural gas from wells to the locations of demand. Substitution of basic steel for wrought-iron in tubes was tried, successfully, in 1884. Pipes and tubes made in the United States in 1880 were lap-welded and once it was known that steel could be welded successfully, the greater durability of this metal guaranteed it would supplant iron eventually in the production of pipes and tubes. Wrought-iron did continue to be used in pipe lines built to convey petroleum and natural gas. By 1890, a technique for producing seamless steel tubing had been perfected.[1] Skelp, a flat bar, was welded into pipe of various sizes.

In addition to transportation related demand for steel and urban related steel use, the demand for steel in manufacturing was expanding during this period. In particular, steel plates and bars were used extensively in heavy equipment.

---

[1]Victor S. Clark, History of Manufactures in the United States, Vol. II, New York: Peter Smith, 1949), p. 347.

B. Geographic Distribution of Demand

As several of the location theorists cited in chapter 2 indicated, location of demand is likely to influence an industry's locational pattern. Thus, to explore the steel industry's optimal location, 1879-1919, either data indicating the geographic distribution of demand must be located or acceptable proxies constructed. The Census Bureau recorded the types of steel products produced (and presumably in demand); unfortunately, neither that agency, nor any other I have found, kept extensive records on the location of steel demand. Therefore, suitable proxies for direct observations of demand locations must be constructed. Due to the limitations of the available data, several somewhat arbitrary assumptions are necessary to obtain a total demand tonnage for each consumption region, as described below.

The first step in constructing demand at various locations stems from the observation that most of the steel output, 1879-1919, was used for one of three purposes: transportation systems, urban development, or manufacturing. From contemporary accounts, I attributed the entire rail tonnage and half of the wire rod tonnage to the transportation category of steel demand. With this assumption and either census or industry data, the tonnage of steel used for transporation can be calculated for each of the study years. Similarly, all structural steel and skelp tonnage, half the wire rod output, and half the tonnage of bars and rods is attributed to urban steel demand to determine urban steel tonnage for each year. Finally, manufacturing demand is calculated as one hundred percent of plate and sheet tonnage, and fifty percent of bars and rods output (excluding wire rods). Table 3.5 shows the results of aggregating steel output in this fashion.

Once the total tonnages for each of the three use categories are derived, the question of where these tonnages were used arises; i.e., what localities had heavy demand for urban steel? for transportation steel? for manufacturing steel? In absence of direct evidence, answering these questions for the transportation demand example requires an estimate of the percentage of total U. S. transport demand accounted for by a particular locality (i) at time j. When that estimated percentage (t_{ij}) is multiplied by total

U. S. transport demand, $(T_{us,j})$, given in tons, an estimated tonnage of transport steel demanded by locality i $(T_{i,j})$ is obtained. Railroad mileages form the data basis from which the percentage are made. In particular, the total U. S. demand for transport steel is assumed to have two subcomponents, the first dependent upon the decadal growth of transportation and the second a function of the current level of transportation facilities. Initially I assumed the growth portion (μ_G) constituted 80% of total U. S. transport steel tonnage and the absolute level of transportation, (μ_A), 20%. This 80%-20% breakdown seemed consistent with contemporary descriptions of growth of service versus maintenance levels of investment in transportation. However, because the 80% growth, 20% absolute level figures were somewhat arbitrary, two other sets of demand calculations were made to test the sensitivity of the estimates of demand locations to the percentages assigned to growth subcomponent versus absolute level subcomponent of transport steel demand.[1] Thus equations (3.1),

---

[1]The other figures were μ_G =50%, μ_A = 50%, μ_G =20%, μ_A = 80%.

Table 3.5

DEMAND FOR STEEL, 1879-1919

Year	Tonnage (2240 lbs.) of Railroad Demand	Tonnage (2240 lbs.) of Urban Demand	Tonnage (2240 lbs.) of Manufacturing Demand
1879	1,217,497	715,289	782,971
1889	2,091,686	1,669,502	1,611,631
1899	2,709,837	3,298,752	3,241,626
1909	4,097,913	6,195,719	5,978,816
1919	3,425,959	7,705,738	10,495,708

Derived from The Census of Manufacturers, 1880, 1890, 1900, 1910, 1920. In this table, no adjustment for net exports of steel has been made. Over the forty year period railroad related demand for steel increased 181.5%; urban demand, 977.3%; and manufacturing demand, 1240.5%.

(3.2), (3.3) are assumed to describe U. S. demand for transportation steel at time j:

(3.1) $T_{us,j} = T^G_{us,j} + T^A_{us,j}$

(3.2) $T^G_{us,j} = \mu_G \, T_{us,j}$

(3.3) $T^A_{us,j} = \mu_A \, T_{us,j}$

Similarly, each locality's demand for transport steel is assumed to have subcomponents. The first, T^G_{ij}, is dependent on the decadal growth of railroad mileage within the locality. The second, T^A_{ij}, is based on total railroad mileage within the locality at time j. To obtain T^G_{ij}, the percentage of total U. S. decadal growth in railroad mileage that occurred within locality i, t^G_{ij}, is multiplied by the U. S. tonnage figure for growth of transportation ($T^G_{us,j}$). The second transport tonnage for locality i, T^A_{ij}, is derived by multiplying the percentage of absolute U. S. transport steel tonnage accounted for by locality i (t^A_{ij}) times the absolute U. S. transport steel tonnage ($T^A_{us,j}$). Each locality's demand in tons for transport steel at time j, T_{ij}, is the sum of these subcomponent demands as indicated in equation (3-4).

(3.4) $T_{ij} = T_{ij}^{G} + T_{ij}^{A}$

or

(3.5) $T_{ij} = t_{ij}^{G} \; T_{us,j}^{G} + t_{ij}^{A} \; T_{us,j}^{A}$

This rather elaborate set of calculations, made for each state or territory in the United States for each year in the study, yields one portion of total steel demand. The urban and manufacturing components remain, but are derived in a manner analagous to that for transport demand.

The urban demand for steel within the U. S. was also assumed to be based upon the growth of urban areas and the absolute size of urban areas. Again, several sets of calculations were made assuming different weights for growth of urban areas, σ_G, versus absolute size of urban areas, σ_A.[1] Thus, total urban steel demand in the United States at time j, $U_{US,j}$, has a growth associated subcomponent, $U_{US,j}^{G}$, and an absolute urban size subcomponent, $U_{US,j}^{A}$, or

(3.6) $U_{us,j} = U_{us,j}^{G} + U_{us,j}^{A}$

---

[1]The weights used were (a) σ_G=.80, σ_A=.20; (b) σ_G=.50, σ_A=.50, (c) σ_G=.20, σ_A=.80.

where

$$(3.7) \quad U^G_{us,j} = \sigma_G U_{us,j}$$

$$(3.8) \quad U^A_{us,j} = \sigma_A U_{us,j}$$

Each locality's portion of U. S. growth associated urban tonnage is that locality's share of U. S. decadal growth in urban population; this share is designated as u^G_{ij}. Each locality's percentage of U. S. tonnage associated with absolute urban size, u^A_{ij}, is that locality's percentage of U. S. urban population at time j.[1] Total demand for urban steel generated by location i at time j therefore becomes

$$(3.9) \quad U_{i,j} = u^G_{ij} U^G_{us,j} + u^A_{ij} U^A_{us,j}$$

Such a total is calculated for each state and territory for each of the years 1879, 1889, 1899, 1909, and 1919.

The third major use of steel involved manufacturing. Total U. S. demand for manufacturing

---

[1]Urban population consists of those in cities having a population of 25,000 or more; these figures are derived from data in the *Census of Population* for 1880, 1890, 1900, 1910, and 1920.

steel ($M_{US,j}$) is divided into the tonnage required to provide for growth of manufacturing ($\varepsilon_G M_{US,j}$) and that required to maintain the absolute level of manufacturing ($\varepsilon_A M_{US,j}$). Three sets of calculations were made for three different assumptions about the size of ε_G and ε_A.[1] Thus, total U. S. manufacturing demand at time j becomes

(3.10) $M_{us,j} = M^G_{us,j} + M^A_{us,j}$

Locality i's percentage (m^G_{ij}) of U. S. growth of manufacturing steel demand is assumed equal to that location's percentage of U. S. decadal growth in horsepower used for manufacturing. Each location's percentage (m^A_{ij}) of U. S. absolute manufacturing demand for steel at time j is assumed equal to that location's percentage of U. S. horsepower used in manufacturing at time j. Thus,

(3.11) $M_{ij} = M^G_{ij} + M^A_{ij}$

(3.12) $M^G_{ij} = m^G_{ij} \; M^G_{us,j}$

---

[1]The calculations were for (a) ε_G=.80, ε_A=.20; (b) ε_G=.50, ε_A=.50; ε_G=.20, ε_A=.80.

(3.13) $M^A_{ij} = m^A_{ij} M^A_{us,j}$

Finally, to obtain each locality's total demand for steel at time j, D_{ij}, the subcomponent parts must be summed, yielding

(3.14) $D_{ij} = T_{ij} + U_{ij} + M_{ij}$

or

(3.15) $D_{ij} = t^G_{ij} (\mu_G T_{us,j}) + t^A_{ij} (\mu_A T_{us,j}) + u^G_{ij} (\sigma_G U_{us,j}) + u^A_{ij} (\sigma_A U_{us,j}) + m^G_{ij} (\varepsilon_G M_{us,j}) + m^A_{ij} (\varepsilon_A M_{us,j})$

In equation (3.15), the $T_{US,j}$, $U_{US,j}$, and $M_{US,j}$ are data originally available by distributing steel output among three use categories; the μ_G, μ_A, σ_G, σ_A, ε_G, and ε_A are given several somewhat arbitrary values; and the t^G_{ij}, t^A_{ij}, u^G_{ij}, u^A_{ij}, m^G_{ij}, and m^A_{ij} are assumed equal in value to their respective proxies.

An additional adjustment to the data must be made to account for exports and imports (negative exports) of steel. Data limitations result in a fairly crude adjustment, but one which does reflect the reduction in demand for domestic steel which imports create around the entry point and the increase in demand for U. S. steel which exports create at the port from which these are shipped. First, imports are added to the total U. S. steel demand tonnages (or exports subtracted), divided equally among the three demand categories. Then the calculations outlined above for determining each locality's demand for steel in tons are performed. Finally, the tonnage of steel imports is subtracted from (or exports added to) the demand figure of east coast localities. It is assumed, and sketchy evidence from the Census Bureau's *Census of Navigation* supports the assumption, that steel imports came primarily through the ports of New York, Philadelphia, and Baltimore or that exports were shipped through these ports. Further, given Census of Navigation reports, New York-Philadelphia ports are assumed to have handled three-quarters of this trade in 1879 and 1889, with Baltimore handling the remaining one-quarter. For 1899, 1909, and 1919, Baltimore and New York-Philadelphia are assumed to have each handled half of the trade.

Eight major consumption regions are formed from the states and territories making up the U. S. market for steel. The eight sites are chosen to reflect patterns of trade indicated by historical accounts of U. S. development. By aggregating the steel demand tonnages for each state/territory within a given region, the steel demand (in tons) for each of these regions and for each year of the study are obtained. Each of the eight regions is assigned a focal point for steel demand in the surrounding region; i.e., it is assumed that steel was shipped to the focal point and then distributed to the surrounding region.[1] The choice of the consumption sites and distribution areas is based on contemporary and historical accounts of the industry and on contemporary transportation systems. The regions are

1. North Atlantic including Maine, New Hampshire, Rhode Island, Vermont, Massachusetts, Connecticut, New York, New Jersey, Delaware, and that part of Pennsylvania which is east of the Allegheniesi, the focal point of this area is New York;

2. South Atlantic, including Virginia, North Carolina, South Carolina, Georgia, and Maryland with Baltimore as the focal point;

---

[1]Limitations on the number of variables permitted by the computer program used to solve the locational model developed in Chapter 4 make narrowing the number of market centers necessary. Although allowing each state or portion of a state to be a market might be preferable, such narrowing does not do severe injustice to the likely market patterns.

3. Gulf, including Florida, Alabama, Mississippi, Louisiana and Texas, with New Orleans as the focal point;

4. Eastern Great Lakes, including Ohio, Michigan, western Pennsylvania, and West Virginia with Cleveland as the focal point;

5. Western Great Lakes, including Indiana, Illinois, Minnesota, Wisconsin, Iowa, North Dakota, and South Dakota, with Chicago as the focal point;

6. Lower Mississippi, including Missouri, Kentucky, Arkansas, Oklahoma, and Tennessee, with St. Louis as the focal point;

7. Rocky Mountain, including Nebraska, Kansas, New Mexico, Wyoming, Montana, and Colorado, with Denver as the focal point;

8. West, including California, Arizona, Utah, Nevada, Idaho, Oregon, and Washington, with San Francisco as the focal point.

Tables 3.6 through 3.8 contain the resulting region demand figures. For each year, the demand figures are given for three different sets of assumptions about the relative contribution of growth versus absolute size to the tonnage of steel demanded within each category (transportation, urban, and manufacturing). Despite the range of weights encompassed in the assumptions, the resulting demand tonnages per region show very little variation.

Table 3.6

GEOGRAPHICAL DISTRIBUTION OF DEMAND FOR THE PRODUCTS OF STEELWORKS AND ROLLING MILLS, 1879-1919, ASSUMING GROWTH IN EACH CATEGORY ACCOUNTS FOR 80% OF TOTAL DEMAND, AND ABSOLUTE SIZE OF PROXY ACCOUNTS FOR 20% OF THE TOTAL

Region	Demand in 1879	Demand in 1889	Demand in 1899	Demand in 1909	Demand in 1919
North Atlantic	846,560	1,333,743	3,391,988	5,747,858	9,969,431
South Atlantic	95,709	181,703	872,541	1,440,504	4,164,485
Gulf	143,894	557,169	668,183	1,602,793	1,020,260
Eastern Great Lakes	469,036	849,497	1,204,449	1,992,219	3,292,814
Western Great Lakes	667,045	1,227,741	1,579,720	2,435,857	2,518,374
Lower Mississippi	220,970	439,866	616,037	1,348,207	716,101
Rocky Mountain	194,014	761,648	336,976	776,781	434,557
West	241,404	409,375	529,172	1,729,051	1,571,843

Derived from *Census of Manufactures*, 1880, 1890, 1900, 1910, 1920; *Census of Population*, 1880, 1890, 1900, 1910, 1920; *Historical Statistics of the United States from Colonial Times to 1957*; *Annual Report of the Interstate Commerce Commission*, 1889, 1899, 1909, 1919.

Table 3.7

GEOGRAPHICAL DISTRIBUTION OF DEMAND FOR THE PRODUCTS OF STEELWORKS AND ROLLING MILLS, 1879-1919, ASSUMING DECADAL GROWTH IN PROXY VARIABLE FOR EACH CATEGORY REFLECTS 50% OF DEMAND IN THAT DEMAND CATEGORY AND THE ABSOLUTE SIZE OF PROXY REFLECTS 50% OF DEMAND IN THAT CATEGORY

Region	Demand in 1879	Demand in 1889	Demand in 1899	Demand in 1909	Demand in 1919
North Atlantic	703,042	1,330,443	3,620,818	5,211,832	8,285,389
South Atlantic	75,883	192,503	858,107	696,895	4,151,361
Gulf	145,222	502,392	753,392	1,460,417	1,002,615
Eastern Great Lakes	468,249	824,650	1,227,052	2,110,668	3,108,891
Western Great Lakes	619,534	1,216,019	1,586,979	2,600,309	2,493,634
Lower Mississippi	219,434	636,810	613,316	1,296,443	906,338
Rocky Mountain	165,914	529,869	381,909	765,719	470,260
West	129,167	429,033	612,308	1,505,557	1,800,462

Table 3.8

GEOGRAPHICAL DISTRIBUTION OF DEMAND FOR THE PRODUCTS OF STEELWORKS AND ROLLING MILLS, 1879-1919, ASSUMING DECADAL GROWTH IN PROXY VARIABLE FOR EACH DEMAND CATEGORY REFLECTS 20% OF DEMAND IN THAT CATEGORY AND THE ABSOLUTE SIZE OF THE PROXY REFLECTS 80% OF DEMAND IN THAT CATEGORY

Region	Demand in 1879	Demand in 1889	Demand in 1899	Demand in 1909	Demand in 1919
North Atlantic	1,093,765	1,590,054	3,592,017	6,127,642	8,601,368
South Atlantic	137,820	203,302	1,089,660	1,398,087	4,135,237
Gulf	167,727	543,868	611,678	1,317,915	984,432
Eastern Great Lakes	546,370	847,912	1,250,742	2,231,189	2,924,966
Western Great Lakes	692,571	1,245,659	1,584,810	2,753,299	2,468,899
Lower Mississippi	261,180	453,088	821,320	1,245,559	857,636
Rocky Mountain	145,569	472,108	851,977	722,171	505,958
West	136,438	334,352	466,585	1,246,407	1,357,808

A surprising feature of all three demand estimates is their strong eastward movement over time; the literature on the late nineteenth century American economy often mentions the westward movement of demand as an important characteristic of economic growth. Even when growth in the proxy variable is assumed to account for 80% of demand in each demand category, 33% of demand is concentrated in the North and South Atlantic regions in 1879; in contrast, about 60% of demand is located in these two regions in 1919. If west coast demand is included (as becomes reasonable with the opening of the Panama Canal), the 1919 east coast figure is 66%. Rising exports and the increased importance of urban related uses of steel contribute heavily to this eastward movement.

C. Cyclical Fluctuations in Aggregate Steel Output

In examining demand for steel at the end of the nineteenth century, the frequent fluctuations during that era in the aggregate level of U. S. output should be kept in mind. Demand for steel, heavily dependent on investment in transportation, buildings, and capital equipment, fluctuated with the level of aggregate economic activity. An interesting question (not pursued here) is whether cyclic variations in aggregate demand influenced the relative location of steel production. During the initial part of the period studied here, 1879 to 1899, downturns in aggregate economic activity probably led to price reductions rather than heavy production cutbacks in the steel industry. After the formation of the United States Steel Corporation in 1901, price competition during severe recessions was less likely, perhaps, than cutbacks in production; the decreased price competition seems probable with the increase in industrial concentration resulting from the formation of the United States Steel Corporation.[1]

As background to the overall level of steel production, the activity of the business cycle between 1880 and 1920 will be briefly traced.

---

[1]This reduced price competition has been discussed in numerous studies on industrial concentration. Walter Adams points out that before the formation of U. S. Steel, the price of Bessemer steel rails fluctuated from $67.52 per ton to $17.62 per ton (1880 to 1901). Within two months of the incorporation of U. S. Steel, the price of Bessemer steel rails settled at $28 per ton and remained there until April, 1916. See Walter Adams, "The Steel Industry" in Walter Adams, ed., <u>The Structure of American Industry</u>, (New York: The Macmillan Company, 1971), pp. 70-116; p. 95.

W. C. Mitchell and A. F. Burns find trough years of the business cycle in 1878, 1885, 1888, 1891, 1894, 1896, and 1908 in the American economy before 1910.[1] Not all these dates represent equally severe downturns. The panic of 1873 and the ensuing depression were both severe and long lived. After a relatively mild downturn in 1882, the 1880's were prosperous, but the panic of 1893 initiated a fairly severe depression; recovery of the economy was on its way by late 1896. There was some stringency in the money market in 1899, accompanied by decline in confidence. Recouperating quickly, the economy continued successfully until the set-back in the years following 1907, but the setback was milder than the recessions following the panics of 1873 and 1893. From 1907 until 1914, business was generally properous. In 1914 and early 1915 the American economy was descending from its prosperous peak but the outbreak of the war checked the downturn. In 1920, there was a

---

[1]Wesley C. Mitchell and Arthur F. Burns, Measuring Business Cycles, (New York: National Bureau of Economic Research, 1947), p. 47, cited in "Editor's Introduction," Ralph Andreano, ed., New Views on American Economic Development, (Cambridge, Massachusetts: Schenkman Publishing Company; 1965), p. 258.

sharp, short depression followed by a long period of prosperity.[1] These cyclic fluctuations are ignored in the model developed in Chapter 4.

---

[1]Clark, History of Manufactures, Vols. II, III, provided the background material on specific cycles.

VI. INDUSTRIAL CONCENTRATION IN THE STEEL INDUSTRY

Coincident with the rapidly rising output of the steel industry was an increase in the concentration of market power within the industry at the turn of the century. This increased industrial concentration was prevalent in American industry around 1900 and often was the result of mergers among previous competitors. Increased industrial concentration and increased geographical concentration do not occur together necessarily; however, if economies of scale in production become extensive relative to market size, both locational and industrial concentration are likely to rise. Further, with an increase in locational concentration, interchange among producers could become easier, facilitating price fixing agreements, etc.

Although local production for a primarily local market had not been the pattern of location in the iron and steel industry since the early nineteenth century production of charcoal pig iron, the other extreme of centralized production for a national market was not typical either.[1] Despite

---

[1]The locational patterns described refer to the "normal" pattern. Producers of specialized products, or serving very isolated markets, existed

the apparent dominance of the Pittsburgh area at the end of the nineteenth century, there was still a pattern of regional production serving primarily, though certainly not exclusively, regional markets.[1] There was competition among producers of different regions, but there was a tendency for the East Coast producers to handle business east of the Alleghenies. Pittsburgh producers shipped steel east and west, though transportation costs generally kept western Pennsylvania steel from being sold on the east coast. The producers bordering the Great Lakes shipped to the growing western market; also, especially for those manufacturers located on Lake Erie, shipment to the eastern New York area was reasonable. Intense competition for growing markets likely led to some crosshauling periodically in the late nineteenth century steel industry, with this tendency aggravated by the cyclic nature of the economy and of the steel market in the late 1890's. Producers of steel were aware of the high fixed

---

in the steel industry throughout the years being discussed.

[1]That the steel industry felt compelled to set up the basing price system to protect Pittsburgh interests suggests the market was not naturally a national one.

costs of production and of the decreasing unit costs that existed over a large range of output. The wide fluctuations in demand, accompanied by a desire to keep average costs down by spreading costs over an efficiently large output, encouraged firms to entice away competitors' customers, especially during slack periods. Attempts to gain new customers could have included shipping into other regions, even if the net price to the producer was fairly low.

The conditions in the steel industry which encouraged reaching beyond a producer's region for new consumers during slack periods also encouraged the formation of pools to fix market shares and prices. Almost invariably these agreements collapsed as soon as the market came under pressure from weakening demand. The Bessemer Steel Association's members, mainly Pennsylvania manufacturers, were frequently indignant at the secret price cuts their comembers gave customers to attract business.[1] Overall, as the _Report of the Commissioner of Corporations on the Steel Industry_ suggests, the steel pools were not very

---

[1]Wall, _Andrew Carnegie_, pp. 330-335.

effective in fixing prices.[1]

In the last half of the nineteenth century, railroads had been major consumers of first iron and later steel. Further, steel production in 1919 was over 32 times greater than it had been in 1879 and was almost 5 times the combined iron and steel output of 1879. As railroad expansion slowed and steel production increased, customers were needed to absorb this steel output; gradually steel moved into these other uses, but the process took time. This transition in the use of steel undoubtedly put even more pressure on producers to fix markets.

Between 1879 and 1919 the average steel-works and rolling mill establishment in the United States increased over fourfold in annual tonnage produced. Pressures had developed for vertical expansion as well as horizontal growth. The integration of blast furnace (pig iron) production and steel production came quickly, as fuel economies inherent in integration were recognized. This integration had occurred previously between blast furnaces and iron rolling mills and the extension

---

[1]United States Commissioner of Corporations, _Report of the Commissioner of Corporations on the Steel Industry_, pp. 68-75.

to steel was logical. Bessemer steel required special qualities of pig iron and apparently the open market was not a dependable source. Related to this and also encouraging vertical combination of blast furnace and converter operations was the large requirement of the Bessemer converter for pig iron relative to the output of the average blast furnace.[1] Only the fuel economies necessarily suggested integration, but the imperfectly competitive nature of both pig iron and steel markets encouraged integration.

The combination of pig iron production and steel production to provide both fuel economies and a guaranteed source of input for the converter or blast furnace occurred quickly in the industry. Most steelworks were also associated with rolling mills.

Integration backwards to materials deposits was not essential for production economies. However the imperfectly competitive nature of first the coke and later the ore market encouraged

---

[1]Temin, Iron and Steel in Nineteenth Century America: An Economic Inquiry, pp. 167-168.

integration by the 1890's.[1] Whether this feeling that control over materials was essential was a totally rational one or not, it seemed widespread. And once a few firms began purchasing large reserves of first fuel and later ore such a policy for the remainder became much more rational. For the fully integrated firm, location of assets was likely to be pretty scattered, since ore, markets, and good fuel only infrequently coexisted. In the regions where this coexistence occurred, such as central Alabama and eastern Pennsylvania, ownership of supplies by steel producers was not unusual. In the early 1890's, however, vertical integration was not very advanced. The U. S. Commissioner of Corporations describes ore production as being a separate business from steel production in the early 1890's. Numerous merchant pig iron furnaces existed and most producers of steel sold their product to finishers of steel. And nearly all iron and steel producers depended upon separate railroad

---

[1]Rockefeller, given his experience in the oil industry, apparently could not understand why, in the early 1890's, steel producers did not buy up the major iron ore reserves. See Hughes, <u>The Vital Few</u>, (London: Oxford University Press, 1965), pp. 257-258.

and water transport companies for transportation of materials and products. The ownership of Pennsylvania coking coal was somewhat more concentrated, since the Carnegie steel interest had merged with the Frick coke interests in the 1880's. During the 1890's, interest in controlling raw materials increased among steel producers. The Illinois Steel Company, for example, acquired both Pennsyvlania coking coal fields and Great Lakes ore property. The Carnegie Steel Company, Ltd., already having secured tremendous reserves of coal, began to acquire ore property.[1] By the late 1890's ownership of raw materials and transportation facilities had become highly concentrated in the hands of semifinished steel producers. In 1890, there were two large concerns in the industry with capitalization around $20,000,000 each: the Illinois Steel Company, with plants located in Chicago, Joliet, and Milwaukee, and the Carnegie Steel Company, all of whose steel production capacity was in the Pittsburgh area. The Report of the Commissioner of Corporations lists other major

---

[1]United States Commission of Corporations, <u>Report of the Commissioner of Corporations on the Steel Industry</u>, pp. 66-68.

producers as including in the Pittsburgh area the Jones and Laughlin interests, comprised of Laughlin & Co. (Ltd.) which operated the Eliza Furnaces and Jones & Laughlins (Ltd.), which operated the American Iron and Steel Works. Another major steel producer was the Lackawanna Iron and Steel Company, formed in 1891 as a merger of the Lackawanna Iron and Coal Company and the Scranton Steel Company; this company provided formidable competition further east, as did the Pennsylvania Steel Company. The latter produced steel outside Harrisburg, Pennsylvania, and had also organized a subsidiary, the Maryland Steel Company. Through its Maryland subsidiary, the Pennsylvania Steel Company had built a plant at Sparrows Point, Maryland, to use imported ore. Two other concerns that were not large absolutely but were big in the special areas where they were located were the Colorado Fuel and Iron Company, near Pueblo, Colorado and the only important producer in the far west, and the Tennessee Coal, Iron, and Railroad Company. The Tennessee concern was a major owner of Southern coal and ore properties and had been an important component in the southern manufacture of pig iron. Both these companies became more involved in steel production in the late 1890's. Two other large Pennsylvania concerns, distinguished by the extent of their operations in the late 1890's, were the Cambria Iron Company and the Bethlehem Iron Company.[1] This competitive situation, with about ten firms accounting for approximately half the industry's output had been typical for the industry since the early 1880's.

The large steel producers described above were involved in the production of crude or semi-finished steel. The producers of finished steel were primarily nonintegrated concerns and highly competitive until the late 1890's when consolidations occurred among producers of many of the finished products (though not generally across product lines).

In 1898 and 1899, several consolidations occurred, leaving three dominant producers of semi-finished steel: (1) the Federal Steel Company, combining the Illinois Steel Company, the Lorain Steel Company, ore mining interests and some railroad property; (2) the National Steel Company, comprising most of the principle crude steel

---

[1] Ibid, pp. 63-65.

facilities west of the Alleghenies other than those held by the Carnegie or Federal Steel Companies, and (3) the Carnegie Company, reorganized in March, 1900, which now formally included the Frick coke interests. These companies controlled, respectively, 15%, 12%, and 18% of U. S. steel ingot production.[1] Additional consolidations occurred among concerns processing semifinished steel. These included the formation of American Tin Plate Company, with an almost complete monopoly of that trade; the National Tube Company, controlling most of the nation's tube production; American Hoop Steel Company, merging the major producers of hoop steel; the American Sheet Steel Company; the American Bridge Company, controlling the bulk of heavy bridge and structural work; and the Shelby Steel Tube Company, with considerable control of seamless tubing production. All the above named consolidations were merged to form United States Steel Corporation in 1901. Several other consolidations, including that forming Republic Iron and Steel Company, also occurred in 1900, but remained outside U. S. Steel. Before the formation of United

---

[1] Ibid, pp. 2-3.

States Steel Corporation, as these various branches attempted to secure complete vertical integration, a battle of the giants for control of steel production was formed. This battle, both among the crude steel producers, and among crude and finished steel producers as each vertically integrated into the other's domain, threatened to overexpand considerably the nation's steel capacity. The desire to avoid the cutthroat competition combined with the enticement of handsome promoter's profits led to further consolidation rather than competition with the formation of United States Steel Company in 1901. This giant concern controlled fully two-thirds of steel ingot manufacture. The company's share of United States pig iron output was 43%, but it held 58% of the nation's steel producing pig iron. Further, the Steel Corporation held shares varying from half to four-fifths of rolled steel production.[1] These figures are consistent with those reported by Warren Nutter, in his study of American monopoly at the beginning of

---

[1]Ibid., p. 12.

the twentieth century.[1] Table 3.9 summarizes the spread of industrial control from the 1890's to 1901 in the United States.

Geographical concentration may have provided more opportunity for increases in industrial concentration than would have geographical dispersion. Did industrial concentration, conversely, lead to an increased geographical concentration? By 1901, roughly 2/3 of the nation's steel producing capacity was concentrated in the United States Steel Corporation. That the location of steel output did not centralize after the formation of U. S. Steel suggests that there were not extensive economies of scale resulting from that merger.[2] The firm fairly quickly decided to expand capacity at Gary, Indiana, Duluth, Minnesota, and Birmingham, Alabama, moves toward locational decentralization. The later introduction of the basing point pricing system can be viewed as an attempt to maintain the competitive position of

---

[1]Warren Nutter, The Extent of Enterprise Monopoly in the United States, 1899-1939, (Chicago, Illinois: University of Chicago Press; 1951), pp. 135-137.

[2]Most scale economies would have shown up at the plant level around this time.

Table 3.9

CONCENTRATION OF U. S. CRUDE STEEL PRODUCTION

Year	Companies	% of Steel Ingot Production Controlled
1898[a]	Carnegie Steel Co., Ltd.	
	Illinois Steel Co.	Approx. 50
	Jones & Laughlin interests	
	Lackawanna Iron and Steel Co.	
	Pennsylvania Steel Co.	
	Cambria Iron Company	
	Bethlehem Iron Company	
1899-1900	Federal Steel Co.	15
	National Steel Co.	12
	Carnegie Co.	18
1901	United States Steel Corporation	66

Taken from United States Commissioner of Corporations, Report of the Commissioner of Corporations on the Steel Industry and Warren Nutter, The Extent of Enterprise Monopoly in the United States, 1899-1939.

Pittsburgh.[1] A monopoly market position allows the firm the luxury of inefficient locations longer than would occur under competition; however, monopoly does not guarantee that the firm will make inefficient locational choices except to the extent that the monopolist limits output to keep prices high. This limitation of output limits expansion and presumably new locations will be harder hit than already existent ones. However, the monopoly firm is not under the same compulsion to make efficient choices that the competitive firm is. The monopoly can use its excess profits to maintain an inefficient facility long after competitive firms would have moved or been competed out of business. Alternatively, however, the monopoly firm might be in a better position than its competitive alternatives to coordination production from numerous locations, phasing out inefficient ones.

---

[1]This means only that location at Pittsburgh was a safe, but not necessarily profit maximizing one. A low cost producer might be better off locating near a (new) large market and receiving substantial phantom freight.

VII. LOCATION OF THE AMERICAN STEEL INDUSTRY, 1879-1919.

Alfred Weber contended that industries would be pulled ever more strongly to raw materials sites over time. This increased raw material orientation would occur because previously ubiquitous materials would become scarce with high demand and consumption of them and similarly, the richer grades of raw materials would be exhausted. Thus, production would be increasingly weight losing.[1] However, Isard suggested that demand has played an increasingly important role in locating steel production. His concern was primarily the diminishing amounts of coal required to produce a ton of steel with technological advance and hence the lesser importance of coal sites in situating iron and steel production. His study emphasized a much longer time period than studied here.[2] Louis Hunter in an

---

[1]Weber, Theory of the Location of Industries, p. 75.

[2]Walter Isard, "Some Locational Factors in the Iron and Steel Industry Since the Early Nineteenth Century," The Journal of Political Economy, LVI (June, 1948), pp. 203-217; Walter Isard and William Capron, "The Future Locational Pattern of Iron and Steep Production in the United States," Journal of Political Economy, LVII, (April, 1949), pp. 118-133.

earlier study of nineteenth century heavy industry also suggested that although Pennsylvania remained the predominant center of iron and steel production, there was a definite westward movement in the industry, with the Chicago area showing the most substantial growth in iron and steel production in the country between 1865 and 1914.[1] Chauncy Harris returned to the early American iron industry to find sensitivity to the market in locational decisions. He pointed out that iron rolling mills and plowworks preceded pig iron production in Pittsburgh, for example. He also described the midwest as being the heart of a large potential market, particularly for agricultural related production (e.g., agricultural implements.)[2] Edgar Hoover, writing primarily on the location of the shoe and leather industry, observed that the influence of transfer costs tended to pull location to markets, raw material

---

[1]Louis Hunter, "The Heavy Industries" in Harold F. Williamson, ed., Growth of the American Economy, (Englewood Cliffs, New Jersey: Prentice Hall, Inc.; 1951), pp. 474-494.

[2]Chauncy D. Harris, "The Market as a Factor in the Localization of Industry," Annals of the Association of American Geographers, Vol. XLIV, No. 4, (December, 1954, pp. 315-348).

sources and transshipment points. The attractions between producers and consumers often made two or more of these points the same. He cited, as an example, Chicago in the late nineteenth century, which was both a transshipment point for Great Lakes iron ore and an increasingly important consumer of steel.[1]

Between 1879 and 1919, shifts in the location of the American steel industry, although not extremely dramatic, did occur. Table 3.10 indicates locational movement in steel production between 1879 and 1919 for the major steel producing states.

Absent from the list is Missouri, an area which generated great hopes for iron production in the midnineteenth century, especially after the discovery of the Pilot Knob iron ores.[2] However, the ores were exhausted sooner than expected and

---

[1]Edgar Hoover, Location Theory and the Shoe and Leather Industry, (Cambridge: Harvard University Press; 1937), p. 59.

[2]Clark, History of Manufactures in the United States, Vol. II, p. 197. The Special Reports on Manufacturing, compiled by the Census Bureau in 1905, ranks Missouri as the nation's sixth largest producer in 1870 of classified iron and steel products. By 1905, Missouri had fallen to eighteenth in rank.

Table 3.10

PERCENTAGE OF U. S. STEEL PRODUCTION BY STATES

State	1879	1889	1899	1904	1909	1914	1919
Illinois	22.1	18.7	13.7	11.4	11.4	7.6	7.7
Indiana	-	.7	.5	.6	3.3	7.1	9.2
Mass.	2.2	2.2	0.9	.8	.6	.4	0.5
New York	8.9	2.4	.7	3.5	4.8	3.2	3.5
Ohio	9.4	9.5	17.0	18.5	20.0	23.3	24.0
Penna.	57.1	63.5	60.2	56.6	51.9	50.6	44.9
All other	.3	3.0	7.0	8.6	8.0	7.3	10.2

Derived from the _Census of Manufacturers_, 1880, 1890, 1900, 1905, 1910, 1915, 1920.

the area could not compete effectively, using Great Lakes ores and Connellsville coke, with more favorable locations. The $75,000 bonus paid to the Vulcan Iron Works of St. Louis by the established steel producers of the Pittsburgh, eastern Pennsylvania and the Chicago area in 1878 to dissuade that company from entering steel rail production further encouraged, but did not basically change, the decline of Missouri as an iron and steel center.[1] Another state absent from the list which might be surprising is Alabama. In the days of Bessemer steel production, Alabama's fairly low grade iron ores did not permit that state to compete effectively in steel production. Alabama's ores did permit production of open-hearth steel, but the state was limited in both markets and monetary capital, the combination of the two being sufficient to stifle her emergence as a steel center during this period. Alabama was becoming increasingly important as an iron producing area. The state's potential for economical steel production was recognized in 1907 by the United States Steel

---

[1]Alderfer and Michl, <u>Economics of American Industry</u>, pp. 64-65.

Corporation when it acquired plants in Birmingham. This movement of U. S. Steel into Alabama eliminated the region's lack of monetary capital, but the slow growth of the region overall perhaps inhibited rapid growth of steel production.[1]

As indicated in Table 3.10 Pennsylvania's share of steel production at first rose, reaching a high of 63.5% in 1889, and then fell steadily thereafter, reaching 44.9% in 1919. No other state took over Pennsylvania's position in the industry. Ohio, however, did more than double her share of the market, going from 9.4% to 24.0% of total production over these forty years. Ohio, with much of her steel producing capacity on or near Lake Erie in the Cleveland area, apparently gained more steel business from the use of Lakes ores and the westward expansion of the market than did Illinois, despite the hopes for Chicago. Chicago was an early steel center, with a major steel concern in the Illinois Steel Company, located in that city (the firm merged into United States Steel Corporation in 1901). And substantial steel output was beginning to develop by the end of this

---

[1]Wall, Andrew Carnegie, p. 347.

period with the emergence of Gary, Indiana, as a steel manufacturing center. The United States Steel Company decided to build a giant plant at that city in 1907, supposedly on the basis of careful consideration of materials and transportation costs.[1] The same company also constructed a modern plant at Duluth, Minnesota, in 1910, perhaps indicating its faith in the profitability of westward movement for the industry.

Although New York was never a dominant manufacturer of steel, it produced a sizeable portion of the United States output in 1879 and had not lost this position entirely in 1919. Interestingly, several producers in New York state had been active in early experimentation with the Bessemer and open-hearth processes.[2] In the earlier years, steel production in New York was attracted to the eastern markets of that state; for example, agricultural machinery had been produced extensively in the Troy area. However, by the late

---

[1]Beckmann, Location Theory, p. 15.

[2]Clark, History of Manufactures in the United States, Vol. II, p. 232. The first permanent Bessemer steel facility was built in Troy, New York, though at the same time works were being established at the Wyandotte Works, near Detroit.

nineteenth century, much of this production had moved to Illinois (harvesters, mowers, threshing machines) and Ohio, Indiana, and Kentucky (plows).[1] The steel production which developed in New York around the turn of the century was more a part of the Great Lakes output than it was akin to the eastern steel production. Between 1900 and 1902 the Lackawanna Iron and Steel Company, a combination of the earlier Lackawanna Company and the Scranton Steel Company, moved its production from Scranton to Buffalo to take advantage of the Lake Erie location.[2]

Massachusetts is out of place with the five other states listed. The state very quickly lagged behind the others in steel output. Massachusetts was never a steel producer of great significance. The iron and steel production of the state was often made by machinery users and producers to supply their own needs or by associated firms. Machinery production generally required a high quality steel. For the high quality requirements crucible steel continued in use in the late nineteenth

---

[1]Ibid., p. 361.

[2]Alderfer and Michl, p. 63.

century. Crucible was especially in demand before widespread production of open-hearth steel. Thus, in 1879, the fairly large figure for Massachusetts steel production may have been largely due to crucible steel production in that state to serve a highly specialized demand.

The portion of the nation's output not produced by these six states gradually increased over the forty years. The areas covered in this "all other" category include, among the larger producers, Alabama, New Jersey, Maryland, the Great Lakes states, and Colorado.

In examining the data in Table 3.10, the dominance of Pennsylvania locationally appears to have lessened, 1879-1919. Further, several other states appear to have increased significantly their geographical share of the steel market. Altogether there are seven observations for each of seven regions (Illinois, Indiana, Massachusetts, New York, Ohio, Pennsylvania, and "all other".) To determine if there was any significant relationship between percentage of United States steel output and the passage of time, that percentage was regressed against time separately for each area.[1] The purpose was to determine if a significantly strong relationship between share of output and time existed to warrant concluding that a particular area's share of the market did change over time.

The results are summarized below, where s is share of the market and t is time. The results for Illinois, Indiana, Massachusetts, Ohio and "All other" are each significant at the .01 level, indicating that the null hypothesis of no correlation can be rejected at that level of significance. The value for r obtained in the equation for Pennsylvania is significant at the 2% level. Only the result obtained for New York is insignificant at a 2% or smaller significance level.[2] This result for New York is not surprising, given the relatively large fluctuations its share of output displayed between 1879 and 1919.

For the purposes of the remainder of this study, seven major feasible production points are

---

[1]Trying to combine cross-sectional and time-series data for this type of test seemed illogical, so seven separate regressions were run.

[2]John Freund, Modern Elementary Statistics, 3rd Edition, (Englewood Cliffs, New Jersey: Prentice-Hall, Inc.; 1967), p. 391.

Table 3.11

SHARE OF OUTPUT AS A FUNCTION OF TIME

State	Regression[1] Equation	r	r^2
Illinois	$S_1 = 3.67-2.39t$	-.95	.90
Indiana	$S_2 = -3.10+1.54t$	.90	.82
Massachusetts	$S_3 = 2.37-0.32t$	-.89	.79
New York	$S_4 = 5.36-.38t$	-.38	.14
Ohio	$S_5 = 6.75+2.66t$	.96	.92
Pennsylvania	$S_6 = 44.87-2.53t$	-.87	.75
All other	$S_7 = .74+1.40t$	.88	.77

[1]Variables are S_i: i^{th} location's share of output
t: time

assumed for the steel industry. They are Pittsburgh, Chicago-Gary, Buffalo, Cleveland, Baltimore (Sparrows Point), Pueblo (south of Denver, Colorado), and Birmingham. These coincide for the most part with the areas listed in Table 3.11. To the extent that they do not, the seven feasible production locations are included because they represent actual or potential geographical diversity in steel production location.

VIII. STEEL AS A WEBER TRANSPORT ORIENTED INDUSTRY

Alfred Weber developed several approaches to the problem of industrial location. The approach most closely associated with his name, and the one which he gave major emphasis, suggests that locations of production are chosen to minimize total transportation costs.[1] Weber called this case, where no factors other than transport cost influenced location, transportation orientation.[2] Weber and his successors did not feel that location would always be transport oriented. Indeed, summarizing Weber and his successors, there are special circumstances under which minimum transportation cost locations will prevail and without these conditions, location will tend to be drawn away from the minimum transport cost sites. The transport orientation will be more likely to prevail if

(1) transport costs are a large fraction of the total costs of the delivered product; this is likely the case when

---

[1]Alfred Weber, Theory of the Location of Industries, pp. 45-94.

[2]Ibid.

bulky inputs and/or output are involved;

(2) the industry does not require large amounts of uniquely trained labor;

(3) being close to demand centers to keep abreast of fashion changes resulting in changed demand for the product is unimportant.

These conditions are descriptive of an industry which processes bulky raw materials into a semifinished or finished producer's durable good. The steel industry has frequently been cited as an example of a Weber transportation oriented industry.[1] For the steel industry, this requires locating where the costs of transporting coal and ore to the production site and steel to the consumer are at a minimum.

Weber described this locational problem in fascinating detail. His basic concern was the individual firm, requiring one or two localized resources for production, and facing a constant demand at a single consumption site. Implicit in the

---

[1]See, for example, Beckmann, Location Theory, p. 15.

analysis was the assumption that average cost was constant regardless of the level of output. Another implicit assumption was the non-substitutability of inputs; that is, a necessary minimum of input A is required per unit of the product and a similiar restriction holds for all inputs. In terms of more recent economic analysis, this requirement means the production function is of the Leontieff fixed input-output coefficient variety.[1] The American steel industry between 1879 and 1919 fits this description well, with ore and coal being the necessary and locationally sensitive inputs. The next chapter contains a model of the steel industry which reflects the transport orientation of this industry.

---

[1]A. A. Walters, "Production and Cost Functions: An Econometric Survey," Econometrica, XXXI, (January-April, 1963), pp. 2-5.

CHAPTER 4: THE LINEAR PROGRAMMING MODEL FOR THE WEBER PROBLEM

I. INTRODUCTION

Alfred Weber suggested the locational figure as a tool for actually locating the minimum transportation cost production site with a transport oriented firm or industry. The locational figure is a polygon with each vertex representing a raw material or market site. If the number of these sites is sufficiently small, rules of geometry can be utilized to find the optimal (minimum transport cost) production site.[1] In a mathematical appendix to Weber's book, Georg Pick both elaborated upon mathematical solution of the Weber problem, via geometry, and proposed, as an alternative to the locational figure, the Varignon frame as an analytical device.[2] The frame is a physical

---

[1] Alfred Weber, Theory of the Location of Industries, tr. by Carl J. Friedrich, (Chicago: University of Chicago Press; 1929), p. 67-75.

[2] Georg Pick, "Mathematical Appendix," in Weber, Theory of the Location of Industries, pp. 227-252.

analogue to the locational figure, which, when properly weighted, will indicate the minimum point within the figure.[1] Neither of these alternatives proves useful in the more interesting locational problems involving numerous input and market sites. Fortunately, the Weber problem with many dimensions is amenable to solution through linear programming.

Linear programming, developed extensively during and after World War II, seeks values of variables which maximize or minimize a linear objective function of those variables, subject to one or more linear constraints.[2] This format describes fairly well the Weber problem. Further, with George Dantzig's development of the simplex method for converging to optimal solutions through an iterative process, the linear programming problem became readily adaptable to computer solution,

---

[1]Ibid., pp. 228-229.

[2]Most microeconomics textbooks have a brief discussion of linear programming; for example, see James M. Henderson and Richard E. Quandt, Microeconomic Theory: A Mathematical Approach, (New York: McGraw-Hill Book Co.; 1958), pp. 75-82. A major summary of the role of linear programming within the realm of linear economics appeared in Robert Dorfman, Paul Samuelson, and Robert Solow, Linear Programming and Economic Analysis (New York: McGraw-Hill Book Company, Inc.; 1958).

without which solution of large problems would be difficult.[1] Formally, linear programming involves finding the n values for x_i such that the objective function

$$(4.1) \quad R = \sum_i a_i x_i \qquad i = 1, \ldots, n$$

is maximized (or minimized). The x_i's must also satisfy m linear constraints, one constraint for each c_j, j=1, . . ., m, of the form

$$(4.2) \quad \sum_i b_{ij} X_i \leq C_j$$

with the direction of the inequality dependent upon the particular problem. Additionally, for the solution of the linear programming problem considered here to be meaningful in an economic sense, the x_i must be nonnegative,

$$(4.3) \quad X_i \geq 0 .$$

Associated with every linear programming problem is a dual problem. If the original problem

---

[1]George B. Dantzig, "Maximization of a Linear Function of Variables Subject to Linear Inequalities", in Tjalling C. Koopmans, ed., <u>Activity Analysis of Production and Allocation</u>, (New York: John Wiley & Sons, Inc., 1951), pp. 339-347.

requires finding values of the n variables x_i which maximize (minimize) the objective function R given by (1), the dual is solved by finding values of the m variables u_j which minimize (maximize) the objective function S given in (4):

$$(4.4) \quad S = \sum_j C_j u_j \qquad j = 1, . . ., m$$

The coefficients of the dual objective function are formed from the right hand side of the original problem's constraints. The dual objective function is minimized (maximized) subject to the n constraints (one for each a_i)

$$(4.5) \quad \sum_j b_{ji} u_j \geq a_i$$

The coefficients appearing in the dual constraints are those of the original constraints but with row and column transposed. The right hand side values of the dual constraints are the coefficients of the objective function in the original problem. The directions of the inequalities in the dual constraints are the reverse of those in the original constraints. The values of the u_j must also be nonnegative,

(4.6) $u_j \geq 0$.

Thus, if solution of the original yields values of the n variables x_i which maximize a linear objective function, R, subject to m constraints, the solution of the dual yields values of the n variables u_j, which minimize a linear objective function, S, subject to n constraints. Further, the maximum value of R is equal to the minimum value of S.

II. LINEAR PROGRAMMING AND THE TRANSPORT ORIENTED INDUSTRY

A frequently cited linear programming problem, the transportation problem, seeks an array of shipments between production and consumption points which minimizes transport costs but satisfies all exogenously given consumer demand without shipments from any plant in excess of production capacity.[1] However, this standard transportation problem is not identical to the Weber problem, despite the common objective of minimizing transportation costs. In the Weber problem, raw materials are shipped to production sites determined, albeit indirectly, by the program, and finished steel is shipped from production to consumption sites. Thus, the Weber problem must be carefully formulated to avoid having unknowns on the right hand side of the constraints. Solution of the problem yields an array of raw materials shipments to alternative feasible production sites and product shipments from these production sites to consumption sites. Actual optimal output

---

[1]Sven Danø, Linear Programming in Industry: Theory and Applications, An Introduction (New York: Springer-Verlag, 1974), pp. 60-66.

levels for each production site can be calculated from the shipment figures.

The problem of minimizing the transport costs of ore, coal, and steel is subject to certain production constraints. It is assumed here that the production function for steel is the fixed coefficient, input-output type, introduced and analyzed by Leontief.[1] Ore and coal are assumed the only locationally important inputs. With an input-output production function for steel, a fixed quantity of coal, α_c, and a fixed quantity of ore, α_r, are required per unit of steel output. Coal cannot be substituted for ore or ore for coal. This production function results in L-shaped isoquants; the values of coal and ore per unit of steel output are the reciprocals of their fixed coefficients in the production function. In addition to satisfying the production function constraint, the shipments of steel must meet or exceed the exogenously given and price inelastic demand for steel at the various consumption locations. Formally, the Weber problem requires the following:

---

[1]Wassily Leontief, The Structure of the American Economy, 1919-1929, (New York: Oxford Press; 1951).

c_{ij}: shipments of coal between the m coal sites (i=1, . . ., m) and the q (j=1, . . ., q) feasible production sites.

r_{lj}: shipments of ore between the n (l=1, . . ., n) ore sites and the q feasible production sites.

s_{jk}: shipments of steel between the (j=1, . . ., q) feasible production sites and the p (k=1, . . ., p) demand sites.

The constants of the problem include

s_k^1 : the exogeneously given demand at consumption site k, with demand given for each k=1, . . ., p.

t_{ij}^c: full cost of transporting a ton of coal from i to j.

t_{lj}^r: full cost of transporting a ton of ore from l to j

t_{jk}^s: full cost of transporting a ton of steel from j to k.

$1/\alpha_c$: output of steel per ton of coal; i.e., the coefficient of coal in the steel production function.

$1/\alpha_r$: quantity of steel produced per ton of ore, i.e., the coefficient of ore in the steel production function.

The objective function is to be minimized by finding appropriate values for the (i•j) + (l•j) + (j•k) variables c_{ij}, r_{lj}, and s_{jk}; the function to be minimized is total transportation cost T, or

$$(4.7) \quad T = \sum_{i=1}^{m} \sum_{j=1}^{q} t_{ij}^{c} c_{ij} + \sum_{l=1}^{m} \sum_{j=1}^{q} t_{lj}^{r} r_{lj} + \sum_{j=1}^{q} \sum_{k=1}^{p} t_{jk}^{s} s_{jk}$$

This objective or preference function is minimized subject to the constraint, first, that all demand is met or

$$(4.8) \quad \sum_{j=1}^{q} S_{jk} \geq S_k^1 \quad \text{for each } k = 1, \ldots, p$$

One such constraint exists for each k. The second constraint requires that shipments from the jth location not exceed production there. The shipments from j to k must be determined by the program, just as indirectly, the production in j^{th} location must be. This constraint is

$$(4.9) \quad \sum_{k=1}^{p} s_{jk} - s_j \leq 0$$

One such constraint exists for every feasible production site. Two further constraints describe the production function as one of the Leontief input-output variety; they specify the minimum amount of coal and ore that must be shipped to produce steel in a given location or

$$(4.10) \quad 1/\alpha_c \sum_{i=1}^{m} c_{ij} - s_j \geq 0$$

$$1/\alpha_r \sum_{l=1}^{n} r_{lj} - s_j \geq 0$$

One such pair of constraints exists for each production site j and defines the production function for that site.

The constraints given in (9) and (10), however, involve an unknown production quantity, s_j, that does not appear in the objective function (7). To avoid using the s_j's, the constraints implied by equations (9) and (10) can be combined as

(4.11) $$1/\alpha_c \sum_{i=1}^{m} c_{ij} - \sum_{k=1}^{p} s_{jk} \geq 0$$

$$1/\alpha_r \sum_{l=1}^{n} r_{lj} - \sum_{k=1}^{p} s_{jk} \geq 0$$

with one such set of constraints for each j, j=1, . . ., q.

Finally, all shipments must be nonnegative, or

(4.12) $c_{ij} \geq 0$; $r_{lj} \geq 0$; $s_{jk} \geq 0$ for all i, j, l, k.

III. AN ILLUSTRATION

A. The Minimum Transport Cost Problem

To clarify the problem, assume there are two coal sites, i=1, 2; two ore sites, l=1,2; two feasible production sites, j=1,2; and there are two consumption sites, k=1,2. The following are given

$S_1^1 = 800;\ S_2^1 = 700$	demand at sites 1,2
$t_{11}^c = 5;\ t_{12}^c = 6;$	transport costs on coal
$t_{21}^c = 7;\ t_{22}^c = 6$	per ton
$t_{11}^r = 3;\ t_{21}^r = 4;$	transport costs on ore
$t_{22}^r = 4;\ t_{22}^r = 3$	per ton
$t_{11}^s = 2;\ t_{12}^s = 3;$	transport costs on steel
$t_{21}^s = 3;\ t_{22}^s = 2$	per ton
$1/\alpha_c = 3;\ 1/\alpha_r = 2$	output of steel per unit of coal and ore inputs.

Solution of this problem means finding the values of the twelve variables c_{11}, c_{12}, c_{21}, c_{22}, r_{11}, r_{12}, r_{21}, r_{22}, s_{11}, s_{12}, s_{21}, s_{22}, such that

$$T = 5c_{11} + 6c_{12} + 7c_{21} + 6c_{22} + 3r_{11} + 4r_{12} + 4r_{21} + 3r_{22} + 2s_{11} + 3s_{12} + 3s_{21} + 2s_{22} \tag{4.13}$$

is minimized, subject to the six regular constraints,

$$(4.14)\quad \begin{aligned} &3c_{11} + 3c_{21} - s_{11} - s_{12} \geq 0 \\ &2r_{11} + 2r_{21} - s_{11} - s_{12} \geq 0 \\ &3c_{12} + 3c_{22} - s_{21} - s_{22} \geq 0 \\ &2r_{12} + 2r_{22} - s_{21} - s_{22} \geq 0 \\ &s_{11} + s_{21} \geq 800 \\ &s_{12} + s_{22} \geq 700 \end{aligned}$$

and the twelve nonegativity constraints on the variable shipments. The minimum value of the function T is 7983.33 and occurs with the following shipments:

$c_{11} = 266.67$	$r_{21} = 0$
$c_{12} = 233.33$	$r_{22} = 350.00$
$c_{21} = 0$	$s_{11} = 800.00$
$c_{22} = 0$	$s_{12} = 0$
$r_{11} = 400.00$	$s_{21} = 0$
$r_{12} = 0$	$s_{22} = 700.00$

These imply production at the first site, S_1, of

$$S_1 = 1/\alpha_c\ (c_{11} + c_{21}) = 3(266.67) = 800.00$$

or

$$S_1 = 1/\alpha_r\ (r_{11} + r_{21}) = 2(400.00) = 800.00$$

and production at the second site of

$$S_2 = 1/\alpha_c\ (c_{12} + c_{22}) = 3(233.33) \approx 700.00$$

or

$$S_2 = 1/\alpha_r\ (r_{12} + r_{22}) = 2(350.00) = 700.00$$

B. The Dual

The dual to the minimum problem given in the illustration above is to find the values of $u_1, u_2, u_3, u_4, u_5, u_6$ that maximize

$$R = o \cdot u_1 + o \cdot u_2 + o \cdot u_3 + o \cdot u_4 + 800 \cdot u_5 + 700 \cdot u_6 \tag{4.15}$$

subject to the twelve constraints

(4.16)

$$3u_1 \leq 5 \qquad 2u_2 \leq 3$$
$$3u_3 \leq 6 \qquad 2u_4 \leq 4$$
$$3u_1 \leq 7 \qquad 2u_2 \leq 4$$
$$3u_3 \leq 6 \qquad 2u_4 \leq 3$$
$$- u_1 - u_2 + u_5 \leq 2$$
$$- u_1 - u_2 + u_6 \leq 3$$
$$- u_3 - u_4 + u_5 \leq 3$$
$$- u_3 - u_4 + u_6 \leq 2.$$

Additionally there are six nonnegativity constraints on the six variables u_i.

The dual seeks to find imputed values of steel which maximize the delivered value of steel. The constraints indicate the imputed values of coal and ore delivered to the production sites and steel delivered to consumption sites, since transportation costs are the only costs considered in the problem. However interesting the mathematical formulation of the dual, it does not add significantly to the economic understanding of the Weber problem and will not be pursued further here. Further, the consideration of transport cost minimization (the primal linear programming problem) is preceded by the assumption that the firm wishes to maximize profits. For the transport oriented firm, this means maximizing the delivered value of steel (the dual problem). Thus, implicitly the analysis begin with the dual maximization problem and moves to the primal minimization problem because of the industry's nature.

IV. APPLICATION OF THE MODEL

A. Computations and Data Sources

Following the format outlined above, the minimum transport cost array of coal, ore, and steel shipments for the American steel industry was obtained for 1879, 1889, 1899, 1909, and 1919. Because the computer program used for the actual calculations will handle a maximum of fifty constraints and one hundred variables, the number of designated materials, production, and consumption sites had to be limited. Based on contemporary accounts of the steel industry and more recent industry studies, two coal sources, three ore sources, seven production sites, and eight consumption points were chosen. The coal sources were Connellsville (outside Pittsburgh) and Birmingham. Contemporary sources indicate that coal sources other than Connellsville-Pocahontas (located in West Virginia) and Birmingham provided about 6% of the coal used around 1900; therefore, excluding these miscellaneous sources seems reasonable. Ore sources were the Great Lakes mines (assumed to be shipped from Cleveland); Birmingham; and foreign (assumed to enter the United States via the port of Baltimore). The feasible production sites included Pittsburgh, Baltimore, Buffalo, Cleveland, Chicago, Birmingham, and Pueblo, Colorado. The latter two were not major producers of steel relative to the others, but did seem to meet a special and potentially growing market. Eight consumption regions were designated, as described in Chapter 3. The Weber problem deals with point to point deliveries of materials and products; therefore, one major city within each consumption region was declared the consumption point. These were the North Atlantic (New York); South Atlantic (Baltimore); Gulf (New Orleans); Lower Mississippi (St. Louis); East Great Lakes (Cleveland); West Great Lakes (Chicago); Rocky Mountain (Denver); and West (San Francisco). Even with these restrictions on feasible locations, the model contains ninety-one variables and twenty-two constraints.

The linear programming model requires for each year data on output per unit of ore input ($1/\alpha_r$), output per unit of coal input ($1/\alpha_c$), and level of demand at each consumption point, as well as transport costs between all resource sites and feasible production sites, and from feasible production sites to consumption sites. Input requirements per unit of output were derived from census data and from Isard's work on resource requirements in the steel industry, as described in Chapter 3.[1]

Solution of the model also requires demand for each year at each of the consumption sites. The demand figures were derived as described in Chapter 3. Finally, transport cost data for each of the five years involved and for each of the ninety-one possible shipments are also essential for obtaining a numerical solution. Such data are difficult to obtain, especially for nineteenth century shipments. Some of the rates per ton between given points were directly available; others had to be projected, based on ton-mile rates for similar areas and transportation distance between the points involved.[2]

---

[1]Census of Manufactures, 1880, 1890, 1900, 1910, 1920; Water Isard, "Some Locational Factors in the Iron and Steel Industry Since the Early Nineteenth Century", _Journal of Political Economy_, LVI, (June, 1948); pp. 203-217.

[2]Sources of transport cost information include the _Census of Manufactures_, 1880; _Census of Transportation_, 1880, 1890; _Report of the Commissioner of Corporations on the Steel Industry_;

The data are collected for each of the years 1879, 1889, 1899, 1909, and 1919; a brief summary of the data is presented in the Appendix. Separate linear programming programs are run for each of the aforementioned years, resulting in an array of raw materials and product shipments for each year.

B. Results Generated by the Model

The levels of shipments indicated by the model for each of the years are given in Table 4.1. From these shipments, the actual levels of output in each of the feasible locations and for each year in the study can be calculated by multiplying $1/\alpha_c$ times coal shipments to the location, or by multiplying $1/\alpha_r$ times ore shipments there, or by summing all shipments of steel from that location to

---

the United States Tariff Commission's Report on Preferential Transportation Rates; Report of the Commissioner of Labor on the Costs of Production of Iron and Steel; Warren, The American Steel Industry, 1850-1970; Congressional Hearings on the Regulation of Railway Rates; Henry Fink, Regulation of Railway Rates on Interstate Freight Traffic; Final Report of the National Waterways Commission; Douglas North, "Ocean Freight Rates and Economic Development," Journal of Economic History, XVIII, (1958), pp. 537-555; Paullin, Atlas of the Historical Geography of the United States; Interstate Commerce Commission, Forty Year Review of Changes in Freight Tariffs. Further information on transport costs is included in the Appendix.

consumption points. These minimum transport cost outputs are given in Table 4.1. The most surprising result generated by the program is the dominant role played by Birmingham in the early years of the study. For comparision, Table 4.2 (the derivation of which is explained later in the chapter) gives actual levels of steel production at each of the seven production sites. As Table 4.2 indicates, Birmingham's position in actual steel production is minimal for most of the years being studied, particularly the early ones. Reasons for this divergence between actual and program results for Birmingham are examined at greater length later in this chapter; the significance of this and other differences between the program generated and actual geographical patterns of output is explored in Chapter 5.

Returning to the program generated output patterns, Birmingham appears as a major producer in each of the five years examined. However, its relative importance diminishes over much of the period. Beginning with 100% of production in 1879, Birmingham's optimal output level falls to 81% in 1889, rises to 87% by 1899, but falls dramatically to 46% by 1909. Another dramatic

Table 4.1

OPTIMAL PRODUCTION OF STEEL AND SHIPMENTS OF COAL, ORE - IN TONS OF 2240 LBS.

	1879	
Steel Production:	Birmingham	2,878,632
	All others	0
Coal Shipments :	Birmingham to Birmingham	8,830,156
	All others	0
Ore Shipments :	Birmingham to Birmingham	6,073,058
	All others	0
	1889	
Steel Production:	Birmingham	4,729,542
	Pittsburgh	1,031,200
Coal Shipments :	Birmingham to Birmingham	9,273,607
	Connellsville to Pittsburgh	2,021,960
Ore Shipments :	Birmingham to Birmingham	7,519,139
	Great Lakes to Pittsburgh	1,639,426
	1899	
Steel Production:	Birmingham	7,994,617
	Cleveland	1,204,490
Coal Shipments :	Birmingham to Birmingham	15,493,441
	Connellsville to Cleveland	2,334,203
Ore Shipments :	Birmingham to Birmingham	13,689,408
	Great Lakes to Cleveland	2,062,411

Table 4.1 Continued

OPTIMAL PRODUCTION OF STEEL AND SHIPMENTS OF COAL, ORE - IN TONS OF 2240 LBS.

	1909	
Steel Production:	Baltimore	7,188,362
	Birmingham	7,892,805
	Cleveland	1,992,217
Coal Shipments :	Connellsville to Baltimore	13,069,742
	Birmingham to Birmingham	14,350,552
	Connellsville to Cleveland	3,622,210
Ore Shipments :	Foreign to Baltimore	21,330,416
	Birmingham to Birmingham	23,420,768
	Gt. Lakes to Cleveland	5,911,613
	1919	
Steel Production:	Birmingham	2,170,888
	Buffalo	13,705,759
	Cleveland	5,811,188
Coal Shipments :	Birmingham to Birmingham	2,837,759
	Connellsville to Buffalo	17,916,000
	Connellsville to Cleveland	7,596,319
Ore Shipments :	Birmingham to Birmingham	3,876,583
	Great Lakes to Buffalo	24,474,544
	Great Lakes to Cleveland	10,377,119

Table 4.2

ACTUAL OUTPUT, STEELWORKS AND ROLLING MILLS
1879-1919 (TONS OF 2240 LBS.)

Production Site	1879	1889	1899	1909	1919
Baltimore	867,039	1,704,210	2,066,228	2,425,390	2,929,420
Birmingham	650	52,205	108,318	299,599	689,109
Buffalo	250,703	240,026	137,981	798,225	936,785
Chicago	400,249	1,020,849	1,911,292	3,051,294	4,307,575
Cleveland	381,069	1,128,013	3,737,497	3,097,426	4,984,114
Pittsburgh	900,986	3,504,102	7,155,896	8,260,886	10,278,670
Pueblo	4,500	75,000	275,000	408,587	502,241

Derived from the Census of Manufactures, 1880, 1890, 1900, 1910, 1920, and Clark, History of Manufactures in the United States, Vol. II.

decline leaves Birmingham with 12% of the national output of steelworks and rolling mill products in 1919.

Only in 1889 does Pittsburgh, the major producer between 1879 and 1919 in reality, appear in the list of optimal production locations. In that year, Pittsburgh uses Connellsville coal and Great Lakes ore to produce about one-quarter of total steel output. Pittsburgh is replaced by Cleveland in 1899; also in that year, the Great Lakes producers begin a position in the optimal production pattern that becomes increasingly dominant over the next twenty years. Baltimore makes a large, but abbreviated, appearance in 1909, using Connellsville coal and foreign ore to produce almost half the production for that year. Cleveland and Buffalo, both Great Lakes producers (on Lake Erie) are dominant producers with Buffalo responsible for a somewhat surprising 62% of the total under the optimal pattern of production in 1919.

Thus, the optimal pattern of production from steelworks and rolling mills generated by the linear programming model suggests a major shift away from Birmingham and towards first Pittsburgh, then the east coast (Baltimore) and Great Lakes (Buffalo, Cleveland) producers, and finally to the Great Lakes producers alone. This move is westward, although less westward than the move of actual output. This westward move occurred during a time when markets were, in percentage terms, moving eastward. Of course, the Great Lakes steel producers were not totally isolated from east coast markets, since relatively inexpensive water transport was available between the two.[1] Nonetheless, this movement of steel production towards the Great Lakes sites suggests that demand patterns did not control the location of steel production.

The important role accorded to Birmingham by the program (as compared with its minimal actual role) raises the question of whether the linear programming model adequately accounts for possible differences between Birmingham and other production sites. For example, contemporary

---

[1]The Interstate Commerce Commission was apparently sensitive to rates on competing transport modes in setting rates for particular railroad routes. For example, see the Digest of Hearings on Railway Rates, 1905, which gives the rates in effect from Erie Railroad points, issued to meet water competition.

studies of the industry suggest that northern steelworks and rolling mill owners in this era feared southern competition, though this competition did not materialize. Two major reasons often were given for the failure of Birmingham to reach its steel producing potential: first, the lack of markets for steel products discouraged the Birmingham industry; second, the inferior quality of southern materials inputs made production there more costly (and the product inferior) than in northern steelworks and rolling mills. If the data used here are accurate, the first argument is fallacious, since the program indicates large Birmingham output even with limited southern markets. The second argument may have considerable vailidity. If steelworks and rolling mills in Birmingham required substantially more ore and coal per ton of output than did steelworks and rolling mills elsewhere, production in Birmingham would have been nonoptimal. Given the frequent references in contemporary accounts to the inferiority of Alabama's coal and ore for steel production as compared with other region's resources, a second linear programming problem was solved. This second program used separate output-input coefficients for Birmingham to reflect the lower productivity of that region's ore and coal. The derivations used in this program were based on census data, and were difficult to derive, since very little of Alabama's pig iron was actually refined into steel. The coefficients are given and their derivation explained in Appendix A. Although the coefficients in every year do indicate lower steel output per ton of input in Alabama than in the rest of the country, the limitations of this data make any results from this "Birmingham adjusted" program very tentative.

Even with higher coal and ore requirements, per ton of steel in the Birmingham region than elsewhere, that region is still the country's minimum transport cost steel producer in 1879, as indicated in Table 4.3. Pittsburgh briefly occupies dominant position in 1889, producing 92% of output; Birmingham produces the remaining 8%. In 1899, Baltimore, using Connellsville coal and foreign ore, produces 47% of output, according to the adjusted optimal program, with Birmingham and Cleveland producing 40% and 13% respectively. Baltimore and Birmingham again share the dominant position in 1909, with Baltimore optimally

Table 4.3

PRODUCTION (IN TONS) OF STEEL UNDER THE MINIMUM TRANSPORT COST MODEL WITH BIRMINGHAM INPUT REQUIREMENTS ADJUSTED

	1879	
Steel Production:	Birmingham	2,642,950
	1889	
Steel Production:	Birmingham	406,587
	Pittsburgh	4,988,875
	1899	
Steel Production:	Baltimore	4,286,093
	Birmingham	3,698,808
	Cleveland	1,197,026
	1909	
Steel Production:	Baltimore	7,139,571
	Birmingham	7,849,540
	Cleveland	2,014,610
	1919	
Steel Production:	Birmingham	2,518,610
	Buffalo	12,952,743
	Cleveland	5,528,803

producing 42%, Birmingham, 46%, and Cleveland, 12%, of the total. By 1919, Great Lakes Production becomes much more important, with Cleveland allocated 26% and Buffalo, 62% of total output. This model has Birmingham producing 12% of total steel in 1919.

Thus, the adjustment of the output-input coefficients to reflect the lower quality of Birmingham's ore and coal resources lessened Birmingham's position in the optimal pattern of steel output. However, the adjustment certainly did not eliminate Birmingham from a very important role in the industry, one much more important than the region actually occupied.

C. Optimal Shipment Patterns, Given the Actual Location of Production

The program described in Section II of this chapter determines optimal shipments of coal, ore, and steel (hence, indirectly, levels of production), given unlimited production capacity at each of the seven feasible production sites. A second program, called the constrained capacity program here to distinguish it from the original program, can be devised to limit implied production at each site to a maximum of the output actually produced there.

The object of this constrained capacity program is to determine the minimum transport cost pattern of ore, coal, and steel shipments, given (1) resource requirements, (2) demand for steel at the various consumption sites, and (3) production capacity at each feasible production location. Capacity is assumed equal to the actual level of production occurring at or near the feasible production site, with the sites being the same seven that are in the original problem. To facilitate calculation of transport costs, all production was assumed to occur at the seven feasible production points. More realistically, each city designated as a production point for the industry is the central point (and likely transshipment center) for a surrounding homogeneous production area. Therefore, included as part of actual production for each of the seven feasible production sites is the output of the surrounding area; the following summary gives the areas whose output is included in the production total for each site: (1) Baltimore - includes Maryland, New Jersey, and that portion of Pennsylvania which is east of the Allegheny Mountains; (2) Birmingham - includes all of Alabama; (3) Buffalo - includes all of New York;

(4) Chicago - includes all of Illinois and all of Indiana; (5) Cleveland - includes all of Ohio; (6) Pittsburgh - includes Western Pennsylvania and all of West Virginia; (7) Pueblo - includes all of Colorado.

These areas include most of the output of steelworks and rolling mills in the United States between 1879 and 1919. Further, they provide regions which are internally fairly homogeneous in their attractiveness for steel production. Thus, Baltimore signifies east coast production; virtually all of Alabama's steelworks and rolling mills were near Birmingham. Comparatively little steel was manufactured in New York before the Buffalo works were established, therefore, the Buffalo figure includes output in all of New York state. The center of Illinois and Indiana steel production was on Lake Michigan with Chicago and, later, Gary, being the major production points; Chicago's output includes that of Illinois and Indiana. Cleveland production encompasses all of Ohio, even though some of Ohio's steelworks and rolling mills were in the southern part of the state. However, Cleveland was Ohio's rapidly growing steel center at this time. Included in Pittsburgh's output is the production of West Virginia, a nearby and similar production area.

Census data on the output of steelworks and rolling mills are used to obtain actual output.[1] For some years (notably 1879 and 1889) data on steel are given separately from information on the nonsteel production of rolling mills. For later years, this steel and iron distinction is not made. Because of this and also because the demand data presented in Chapter 3 are based on output of steelworks and rolling mills, it is appropriate to include all steelworks and rolling mills' output in the capacity figures. Using census data and interpolations of census data, the actual output levels indicated in Table 4.2 were obtained.

The linear program for the constrained capacity problem adds seven additional constraints to the program given in Section II. One new constraint exists for each production site and implies

---

[1]In several cases, census data are not directly available. For example, if only a few firms are located within the state, disclosure laws prohibit publishing output and other industry information. In all these cases, either Clark, <u>History of Manufactures</u>, Vols. I, II, had the information or it could be derived from furnace and converter capacity figures given in the census volumes.

that the sum of all shipments from that site must be less than or equal to actual production, q_j, at that site; i.e.,

$$(4.17) \quad \sum_{k=1}^{p} s_{jk} \leq q_j$$

For the problem of minimum transport cost shipments with seven feasible production sites and eight consumption points, there are seven such constraints and within each constraint, eight possible shipments of steel to consumption sites. These eight shipments cannot exceed the site's capacity. Again, with two coal sources, three ore sources, seven feasible production sites, and eight consumption sites, the objective function involves ninety-one possible shipments and therefore contains ninety-one transport cost coefficients. With the seven additional capacity constraints, this program has ninety-one variables and twenty-nine constraints.

Table 4.4 summarizes the results of the constrained capacity program by giving the minimum transport cost shipments of steel to the various consumption sites. Of course, the pattern of shipments is much more diverse than in the original program. The shipment patterns are interesting and, in some cases, surprising. In 1879, Baltimore

Table 4.4

PATTERN OF STEEL SHIPMENTS GENERATED BY THE CAPACITY CONSTRAINED PROGRAM, 1879-1919

1879	From						
To	Baltimore	Birmingham	Buffalo	Chicago	Cleveland	Pittsburgh	Pueblo
N. Atlantic	651,947		87,061				
S. Atlantic	57,346						
Gulf		650				141,243	
L.Mississippi						220,035	
E.Gt.Lakes			163,642			308,835	
W.Gt.Lakes				59,017	381,069	230,873	
Rocky Mtn.				196,309			
West				144,923			
1889							
N.Atlantic			240,026		649,506	56,631	
S.Atlantic						190,194	

Table 4.4 Continued

1899 To	From Baltimore	Birmingham	Buffalo	Chicago	Cleveland	Pittsburgh	Pueblo
Gulf						554,417	
L.Mississippi						443,074	
E.Gt.Lakes						817,018	
W.Gt.Lakes						1,442,768	
Rocky Mtn.				471,116	124,125		
West		52,205			354,382		

1899	Baltimore	Birmingham	Buffalo	Chicago	Cleveland	Pittsburgh	Pueblo
N.Atlantic	2,066,228		137,981			1,217,674	
S.Atlantic						864,210	
Gulf		108,318			557,845		
L.Mississippi					620,865		
E.Gt.Lakes					1,058,299	138,727	
W.Gt.Lakes				1,052,261	500,488		

Table 4.4 Continued

1899	From						
To	Baltimore	Birmingham	Buffalo	Chicago	Cleveland	Pittsburgh	Pueblo
Rocky Mtn.				334,193			
West				524,838			
1909							
N.Atlantic	1,043,433		798,225			3,915,953	
S.Atlantic	1,381,956						
Gulf		299,599				1,288,742	
L.Mississippi					1,359,172		
E.Gt.Lakes						2,014,614	
W.Gt.Lakes				601,063	1,738,253	112,503	
Rocky Mtn.				761,822			
West				1,688,408			
1919							
N.Atlantic						7,195,649	

Table 4.4 Continued

1919	From						
To	Baltimore	Birmingham	Buffalo	Chicago	Cleveland	Pittsburgh	Pueblo
S.Atlantic	1,090,253					3,083,020	
Gulf		689,109			487,826		
L.Mississippi				806,267			
E.Gt.Lakes					3,007,709		
W.Gt.Lakes				1,838,782	682,311		
Rocky Mtn.				535,445			
West	647,034		936,785				

production is used mainly to supply North Atlantic demand, although Baltimore also fully handles South Atlantic demand. Birmingham's limited output in 1879 is sent to the Gulf region; Pittsburgh handles most of the Gulf demand. Buffalo's steel output is shipped primarily to the East Great Lakes area while the East Great Lakes production center, Cleveland, ships its steel to the West Great Lakes region. The portion of North Atlantic demand not handled by east coast suppliers is covered by Buffalo. Most of the western market (including West Great Lakes, Rocky Mountain and West regions) is supplied by Chicago. Pittsburgh handles most of the Gulf, Lower Mississippi, and East Great Lakes demand, as well as a substantial part of the West Great Lakes market. Pueblo has no production under the constrained capacity program, consistent with reports that transport rates for the Colorado producer were set high enough to allow Chicago producers to compete effectively between the Mississippi River and the west coast.[1]

In 1889, the program suggests that the

---

[1]Warren, American Steel Industry, 1850-1970, p. 82.

Atlantic coast demand is supplied from Buffalo, Cleveland, and Pittsburgh. All of Birmingham's output is shipped to the West, and all of Chicago's to the Rocky Mountain market. Pittsburgh handles the bulk of east coast and central U. S. demand.

Cleveland displaces Pittsburgh as a major central U. S. supplier in 1899, with the latter's output shipped only to the North Atlantic and the East Great Lakes regions. Baltimore ships all its production to the North Atlantic, while the South Atlantic market is handled by Buffalo and Pittsburgh rather than Baltimore. The Gulf market absorbs all of Birmingham's production, with the remainder of that region's demand supplied by Cleveland. Chicago again handles the West Great Lakes and Western markets, precluding Pueblo from even the Rocky Mountain market.

In 1909, Baltimore is again fully supplying the South Atlantic market and sending its surplus output to the North Atlantic. Birmingham sends its full production to the Gulf market with the remaining Gulf demand covered by Pittsburgh. Pittsburgh sends about half its output to the North Atlantic, while Buffalo ships its entire output to that market. Again, Chicago handles the entire west coast and Rocky Mountain markets with its residual output going to the nearby West Great Lakes area. The portion of that market not covered by Chicago is supplied primarily by Cleveland with a small shipment from Pittsburgh. Cleveland fully supplies the Lower Mississippi, while Pittsburgh handles the East Great Lakes market (centered on Cleveland).

By 1919, the Panama Canal was in operation, although its official opening was in 1920. The shipments of steel from Baltimore and Buffalo to the west coast undoubtedly reflect the availability of this water route; Chicago, preempted from the west coast market, ships to the Lower Mississippi, Rocky Mountain, and West Great Lakes markets. Baltimore shares the South Atlantic market with Pittsburgh. Birmingham sells its entire output to the Gulf market with Cleveland providing the remainder of that region's needs. Cleveland also handles the entire East Great Lakes market and ships its residual output to the West Great Lakes area. Pittsburgh sends most of its steel to the North Atlantic.

The capacity constrained program indicates excess capacity in a few years. This excess occurs because the demands that must be met by steel shipments are only estimates of true demand. The estimates include almost all steel production, but exclude a miscellaneous tonnage that could not readily be categorized as related to urban, transportation or manufacturing development.

Over the forty year period, the shipment pattern fluctuates considerably. However, some general patterns appear to evolve. Baltimore moves towards handling east and (with the opening of the Panama Canal) west coast demand. Baltimore's output is insufficient to handle the very large North Atlantic demand; the residual of this market is covered increasingly by Pittsburgh, until it handles the entire market in 1919. Pittsburgh also becomes more involved in the South Atlantic market (except in 1909). Chicago produces most of the steel consumed west of the Mississippi, until the opening of the Panama Canal. It also increasingly handles the West Great Lakes market. The Lower Mississippi, East Great Lakes, and West Great Lakes markets are bounced among Chicago, Cleveland, and Pittsburgh suppliers. Most of Birmingham's output is sent to the Gulf region, and is in every year insufficient to meet that demand.[1] Pueblo in no year is a minimum transport cost producer, even for Rocky Mountain area consumers.

---

[1]This may again suggest that the retarded development of the Birmingham steel industry was not entirely caused by inadequate southern demand.

CHAPTER 5: INTERPRETATION OF RESULTS

I. Introduction

In this chapter, the intriguing results of the linear programming model and their divergence from actual output patterns are scrutinized. After a brief review of these patterns, Section II describes three statistical tests applied to the data to determine the statistical significance of the divergence between program generated and actual output levels. The same section summarizes the results of the tests. Section III provides a qualitative analysis of the divergence between the two locational patterns of the steel industry. Finally, Section IV contains concluding remarks on the additional evidence this dissertation's approach has provided for the study of industrial location within a historical context.

A brief review of optimal (program generated) output levels in the seven feasible production sites as compared with actual output levels provides the setting for the remainder of this chapter. Actual output was produced primarily at Pittsburgh over the entire forty year interval;

however, Pittsburgh's role became smaller over time as the Great Lake's producers (Cleveland and Chicago-Gary) grew increasingly dominant. This shift in actual output constituted a westward move, despite the eastward move of markets. The actual shift in the geographical pattern of production also represents a movement away from coal and towards ore supplies.

The optimal locational distribution of steel production is dominated by Birmingham in the early years of the study, in contrast to its minimal actual role. The linear programming results suggest an initial and limited shift from Birmingham to Pittsburgh and then Baltimore. By the turn of the century, however, the program results imply movement towards the Great Lakes production sites. Although these results give Buffalo and Cleveland (Lake Erie sites) preeminent positions, and actual output moved towards Cleveland and Chicago-Gary, the movement in both actual and optimal patterns is in the same direction. Both move in a northwestward direction, after apparently large differences between the two in the early years of the study. The statistical significance of the difference between the optimal and actual location patterns, for individual years and over the entire period, is now examined.

II. Statistical Analysis of the Divergence Between Optimal and Actual Output

A. Nonparametric Tests

Several statistical tests are applied to the data to determine the statistical significance of the difference (D_i) between program generated (Q_i) and actual levels of output (A_i). The tests do not provide irrefutable evidence, but they do suggest both that a difference exists between optimal and actual output initially, and that the difference becomes smaller over time. The first tests are nonparametric. The third is based on regression analysis.

The Wilcoxon matched pairs signed ranks test is applied first. Siegel describes this test as being more powerful than other rank-sign tests because it utilizes information on the magnitude as well as the direction of differences within pairs.[1] With the Wilcoxon test, the D_i's are obtained and ranked without regard to sign, with

---

[1]Sidney Siegel, Nonparametric Statistics for the Behavioral Sciences (New York: McGraw-Hill Book Company, Inc., 1956), pp. 75-83; Siegel's description of the test is followed closely here.

smaller absolute differences having a lower numerical rank than larger absolute differences. If the hypothesis (H_o) that the actual and optimal levels of output are the same is true, then some of the larger differences will favor the actual output level sample and others will favor the optimal output level sample, which implies that some of the larger differences will be positive and some will be negative. If, however, the sum of the negative ranks greatly exceeds the sum of the positive ranks or vice versa, H_o, the hypothesis of no difference between the two samples, would have to be rejected. To determine whether the null hypothesis, H_o, can be accepted or not, T is obtained, where T is the sum of either the ranks having positive sign or the ranks having negative sign, whichever is smaller. Using T tables, an observed value of T equal to or less than the value given in the table for a particular significance level and sample size would indicate that the null hypothesis could be rejected at that significance level. The T tables given by Siegel can be adapted for both one-tailed and two-tailed tests.

Similarly, the Wilcoxon test can be used to test whether the differences (D_i) are getting larger, smaller, or staying the same over time. That is, the seven D_i for 1879 constitute one sample which is compared with the D_i for 1889 in the manner described above. In like fashion, the D_i sample for 1889 is compared with that of 1899; the D_i sample for 1899 with that of 1909; and the D_i sample of 1909 with that for 1919. Finally, the D_i sample of 1879 is compared with that for 1919.

Table 5.1 shows the rank and sign of the differences between optimal and actual output levels for each feasible production site and for each year. The null hypothesis is that the two samples come from the same population. With a two-tailed test, this null hypothesis cannot be rejected at the .05 significance level for any of the years with a sample of seven pairs of observations.[1] Thus, despite the apparent gap between the optimal and actual output levels, the gap is not sufficient with the small sample size for rejecting the null hypothesis. Also, for

---

[1]Siegel, Nonparametric Statistics, p. 254.

Table 5.1

RANK AND SIGN OF DIFFERENCES BETWEEN OPTIMAL AND ACTUAL OUTPUT LEVELS AND T SCORES

	YEAR				
Site	1879	1889	1899	1909	1919
Baltimore	-5	-5	-5	+5	-4
Birmingham	+7	+7	+7	+6	+3
Buffalo	-2	-2	-1	-2	+7
Chicago	-4	-3	-4	-4	-5
Cleveland	-3	-4	-3	-3	+2
Pittsburgh	-6	-6	-6	-7	-6
Pueblo	-1	-1	-2	-1	-1
T	7	7	7	11	12

several of the years, only the Birmingham and Pittsburgh optimal output levels are far different from their actual levels.

A second set of Wilcoxon tests concerns the sample of differences over time; that is, it tests whether the differences in one year are significantly larger than the differences in later years. This one-tailed test of whether differences are becoming smaller over time is used to compare differences between 1879 and 1889, 1889 and 1899, 1899 and 1909, and 1909 and 1919. The sign and rank of the differences for each of these comparisions and for an 1879-1919 comparision, as well as the value of T for each test, are given in Table 5.2. Given the sample size and value of T, the null hypothesis cannot be rejected at the .05 significance level for any of the decadal comparisions.[1] Similarly, the forty year comparison between differences in 1879 and those of 1919, suggests the null hypothesis cannot be rejected for T=10 at the .05 significance level.[2]

---

[1]Ibid.

[2]Ibid.

Table 5.2

A. RANK AND SIGN OF DECADAL DIFFERENCES IN OPTIMAL AND ACTUAL OUTPUT LEVELS AND VALUE OF T FOR EACH TEST

Site	1879-1889	1889-1899	1899-1909	1909-1919
Baltimore	+5	+3	-7	+6
Birmingham	-7	-6	+2	+5
Buffalo	-1	-1	+4	+7
Chicago	+3	+5	+6	+2
Cleveland	+4	+4	-3	-3
Pittsburgh	+6	+7	+5	+4
Pueblo	+2	+2	+1	+1
T	8	7	10	3

Table 5.2 Continued

B. RANK AND SIGN OF DIFFERENCES IN OPTIMAL AND ACTUAL OUTPUT LEVELS AND VALUE OF T, 1879-1919

Site	1879-1919
Baltimore	+4
Birmingham	+2
Buffalo	-7
Chicago	+5
Cleveland	-3
Pittsburgh	+6
Pueblo	+1
T	10

When the individual year observations are combined to provide a sample with thirty-five observations, however, the value of T is 44; this value is significant at the .01 confidence level. Thus, although the Wilcoxon matched-pairs signed-ranks test does not permit rejection of the null hypothesis of no difference between actual and program generated output levels in the small sample cases, it does permit rejection of this hypothesis when more observations are available.

The second nonparametric used is the Kolmogorov-Smirnov two sample test to determine if two independent samples come from the same population.[1] This test compares the cumulative distributions of the two samples at appropriate intervals; if the maximum difference between the two distributions is sufficiently large, the null hypothesis, that both samples are from the same population, can be rejected. To apply the test, cumulative step functions are set up for each sample; the difference between the two samples' cumulative distribution values is calculated for each interval. For a two-tailed test, that difference which has

---

[1]Ibid., p. 127; this description of the Kolmogorov-Smirnov test draws heavily from Siegel.

the largest absolute value, D, provides the basis of the test; the distribution of D has been formulated and tables are available on the probabilities of various sizes of D occurring, given that both samples come from the same population.[1] The test is applied for each year of the study (with seven observations for each sample per year) and also over the entire time period (thirty-five observations per sample). The values of the calculated K_D, the numerator of D, are given for each case in Table 5.3. Also shown are the minimum values of K_D for which the null hypothesis (that the two sets of observations come from the same population) can be rejected, given the number of observations per sample. For the single year samples, the null hypothesis can be rejected only in 1879. Even though the K_D is large in many years, the sample size each year is small and rejection of the null hypothesis at the .05 significance level is not possible. However, with the pooled data, which provide thirty-five observations, the null hypothesis can be rejected at the .05 significance level.

---

[1]Ibid., p. 128.

Table 5.3

CALCULATED VALUES OF K_D AND CRITICAL VALUES OF K_D

AT THE .05 SIGNIFICANCE LEVEL

Sample Year	Sample Size	Calculated Value of K_D	Critical (Tabled) Value of K_D at .05 Significance Level[1]
1879	7	6	6
1889	7	5	6
1899	7	5	6
1909	7	4	6
1919	7	4	6
Combined sample	35	23	12

[1]These values are taken from Table L, p. 278, of Siegel, *Nonparametric Statistics*.

B. Parametric Tests

In addition to the nonparametric tests, two regression equations are estimated and then compared to ascertain the similarity between program simulated and actual levels of steel output. The first regression equation, given in equation (5.1), relates program generated output levels to the transport cost variables of the linear programming model. Equations of this and related forms were estimated for each of the sample years and for the five years' data combined. In addition to the simulated output obtained from the linear programming model using transport costs, several linear programs were run using hypothetical data. One such simulation generated output in various locations with coal transport costs rising fifty percent, all other transport costs remaining constant. A second simulation assumed a fifty percent rise in ore transport costs; a third assumed that steel transport costs were fifty percent higher than they were in reality. These simulations, when combined with the original linear programming results, provide four observations per production site in each year, or twenty-eight observations per cross-sectional sample (one hundred and forty observations for the combined time series-cross sectional sample).

$$(5.1)\quad Q = B_0 + B_1 t_c + B_2 t_r + \sum_{k=3}^{10} B_k t_k + B_{11} T$$

where t_c: coal transport costs to the jth production site per ton of steel output

t_r: ore transport costs to the jth production site per ton of steel output

t_k: transport costs per ton of steel from j^{th} production site to consumption site k

T: dummy variable for time

A number of variations of equation (5.1) were tried and the best fit was obtained when either output or market share was regressed upon all ten transport cost variables (for individual cross-sectional samples; in these samples, the dummy variable T was deleted as an independent variable). For the cross-sectional studies the multiple correlation coefficients ranged from .84 to .96. Unfortunately, a relationship with eleven variables is not very helpful statistically in examining actual output as a function of the transport cost variables. With eleven variables and only seven observations on actual output for each cross sectional sample, regression analysis is rather meaningless. Therefore, equation (5.2) was estimated for optimal and actual data; instead of using transport costs to all markets, (5.2) includes only the cost to the nearest market.

$$(5.2) \quad Q = B_0 + B_1 t_c + B_2 t_r + B_3 t_n$$

where t_c: transport costs of coal per ton of steel to production site j

t_r: ore transport costs per ton on steel to production site j

t_n: costs of transporting a ton of steel to the market nearest production site j

The estimated regression equations for actual and optimal output levels are shown in Table 5.4 for each sample year, as are the corresponding multiple correlation coefficients. Each estimated coefficient for actual output ($\hat{B}_i$) is then compared with the corresponding coefficient for program

Table 5.4

REGRESSION EQUATIONS FOR ACTUAL AND PROGRAM GENERATED OUTPUT AS FUNCTIONS OF TRANSPORT COSTS

Year	Dependent Variable	Coefficients of Dependent Variables t values in parentheses				r	Sample Size
		X_1*	X_2*	X_3*	Intercept		
1879	Q_A	-2.569 (-2.265)	3.263 (1.537)	-21.770 (-2.584)	89.580	.8824	7
	Q	-.082 (-.104)	-.552 (-.370)	45.992 (5.542)	-28.625	.7843	28
	M_A	-.009 (-2.257)	.012 (1.535)	-.078 (-2.580)	.319	.7774	7
	M	-.003 (-.104)	-.002 (-.370)	.174 (5.540)	-.108	.6149	28
1889	Q_A	-13.746 (-1.532)	20.115 (1.423)	-87.133 (-1.620)	245.826	.7968	7
	Q	-8.142 (-3.185)	-.982 (-.243)	116.292 (6.283)	15.576	.8332	28
	M_A	-.018 (-1.535)	.026 (1.426)	-.113 (-1.618)	.318	.7969	7
	M	-.015 (-3.185)	-.002 (-.243)	.216 (6.284)	.029	.8333	28

Table 5.4 Continued

REGRESSION EQUATIONS FOR ACTUAL AND PROGRAM GENERATED OUTPUT AS FUNCTIONS OF TRANSPORT COSTS

Year	Dependent Variable	Coefficients of Dependent Variables t values in parentheses				r	Sample Size
		X_1*	X_2*	X_3*	Intercept		
1899	Q_A	-126.491 (-5.229)	113.427 (4.371)	-271.187 (-3.966)	633.453	.9583	7
	Q	7.087 (.481)	-19.179 (1.201)	229.974 (3.883)	-80.348	.6801	28
	M_A	-.082 (-5.237)	.074 (4.379)	-.176 (-3.970)	.412	.9584	7
	M	.008 (.481)	-.021 (-1.202)	.251 (3.884)	-.088	.6802	28
1909	Q_A	-197.129 (-1.860)	62.439 (1.465)	-242.790 (-1.629)	631.695	.7802	7
	Q	41.852 (1.290)	-34.040 (-2.479)	130.111 (1.852)	217.867	.5972	28
	M_A	-.108 (-1.866)	.034 (1.469)	-0.132 (-1.634)	.344	.7812	7

Table 5.4

REGRESSION EQUATIONS FOR ACTUAL AND PROGRAM GENERATED OUTPUT AS FUNCTIONS OF TRANSPORT COSTS

		Coefficients of Dependent Variables t values in parentheses					
Year	Dependent Variable	X_1*	X_2*	X_3*	Intercept	r	Sample Size
	M	.025 (1.290)	-.020 (-2.479)	.077 (1.852)	.128	.5971	28
1919	Q_A	-119.144 (-1.672)	73.976 (1.241)	-167.975 (-1.356)	884.612	.7403	7
	Q	-12.889 (-.468)	-12.712 (-.545)	24.382 (.388)	398.989	.3726	28
	M_A	-.048 (-1.670)	.030 (1.239)	-.068 (-1.355)	.359	.7402	7
	M	-.006 (-.467)	-.006 (-.546)	.012 (.389)	.189	.3725	28

*Explanation of variables:
Q_A: Actual output
Q : Program generated output
M_A: Actual market share
M : Program generated market share
X_1: Coal transport costs
X_2: Ore transport costs
X_3: Steel transport costs, to nearest market

generated output levels by examining the t statistic defined in equation (5.3).

$$(5.3) \quad t = \frac{\hat{B}_i - B_i}{S_{\hat{B}i}}$$

where $S_{\hat{B}i}$: standard error of the regression coefficient B_i

Table 5.5 summarizes these t values and indicates those which are significant at the .05 confidence level (indicating a significant difference between coefficients).

Examination of the coefficients of the transport cost variables in these regression equations provides some unexpected results. In almost all cases, the coefficient of the coal transport costs variable is negative as expected. For ore transport costs, the coefficient in the minimum transport cost output regression is negative, but it is positive for most actual output regressions. The coefficient of steel transport costs to the nearest market is positive for program generated output as the dependent variable, and negative when actual output is the dependent variable. More complete specification of transport costs might have eliminated some or all of these peculiarities.

Table 5.5 indicates the results of comparing the coefficients ($\hat{B}_i$) in the actual output regressions with the theoretical B_i from the regressions relating optimal output to transport costs. These comparisions were made first with quantity of output as the dependent variable and then with market share as the dependent variable. A "no" in the appropriate space indicates that there is insufficient evidence to reject the hypothesis that $\hat{B}_i = B_i$. For all years except 1919 the test suggests that the coefficients of X_3 (transport cost to the nearest market) differed significantly (given their opposite signs in most cases, this seems reasonable). For every pair of coefficients in 1899, the difference is significant at the .05 confidence level, while for 1919, the difference is not significant for any pair of coefficients. The difference between the coefficients for ore transport costs is significant only in 1899 and the same is true for the coal transport cost coefficient.

Given the sample sizes (7 per cross-sectional sample of actual output; 28 per cross-sectional sample of program generated output levels),

Table 5.5

SIGNIFICANCE AT .05 CONFIDENCE LEVEL OF THE DIFFERENCE BETWEEN COEFFICIENTS OF ACTUAL OUTPUT REGRESSIONS AND OPTIMAL OUTPUT REGRESSIONS (VALUES OF t STATISTIC FOR DIFFERENCE IN PARENTHESES)

Year	Dependent Variable	X_1*	X_2*	X_3*
1879	Quantity of output	no (1.337)	no (1.193)	yes (4.982)
	Market share	no (2.143)	no (1.842)	yes (8.355)
1889	Quantity of output	no (.635)	no (1.492)	yes (3.781)
	Market share	no (.258)	no (1.529)	yes (4.722)
1899	Quantity of output	yes (5.521)	yes (5.110)	yes (7.329)
	Market share	yes (5.732)	yes (3.147)	yes (9.624)
1909	Quantity of output	no (2.255)	no (2.264)	yes (2.501)
	Market share	no (2.308)	no (2.330)	yes (2.578)

Table 5.5 Continued

SIGNIFICANCE AT .05 CONFIDENCE LEVEL OF THE DIFFERENCE BETWEEN COEFFICIENTS OF ACTUAL OUTPUT REGRESSIONS AND OPTIMAL OUTPUT REGRESSIONS (VALUES OF t STATISTIC FOR DIFFERENCE IN PARENTHESES)

Year	Dependent Variable	X_1*	X_2*	X_3*
1919	Quantity of output	no (1.491)	no (1.454)	no (1.553)
	Market share	no (1.451)	no (1.486)	no (1.590)

Variables - X_1: Coal transport cost
X_2*: Ore transport cost
X_3*: Transport cost to nearest market

the F values for the regression equations are significant at the .05 significance level for 1879, 1889, 1899 and 1909 for program generated output. The F values are significant at the .05 level only in 1899 for actual output levels.

Similar regressions are attempted for the combined samples data; the regression variables involving transport costs were less successful in explaining output variations over time than for individual years. The level of transport costs fell between 1879 and 1909; rates rose slightly between 1909 and 1919 but not to 1879 levels. To adjust for these fluctuations in overall transport rates, indexed transport rates were used as independent variables. However, this regression for program generated output levels had about the same explanatory value as unadjusted transport rates plus a dummy variable for time when market share (rather than output tonnage) was the dependent variable. This function with all transport costs as independent variables generated a multiple correlation coefficient of .62. The results when only three transport cost variables appear as independent variables are less attractive. The coefficients of the regression equations using only three transport cost variables, t_c, t_r, t_n, are shown in Table 5.6. Again, the coefficients of the regression equation for actual output are compared with the values of the corresponding coefficients from the regression with optimal output as the dependent variable. The t values and their significance are shown in Table 5.7.

Obviously, in using only the cost of transporting steel to the nearest market rather than including transport costs to all markets in the regression analysis, much of the interdependence provided by the linear programming model is lost, both for individual year samples and for the combined data. Output at a given location probably became increasingly sensitive to steel transport costs to a larger number of markets over time; as transport costs fell, producers became better able to compete in distant markets. Over time, therefore, transport costs to only the nearest market become increasingly inadequate in explaining variations in output. This might explain the relatively poor fits this regression equation provides for the latter years of the study.

In the combined samples regression, the

Table 5.6

REGRESSION EQUATIONS FOR ACTUAL AND PROGRAM GENERATED OUTPUT AS FUNCTIONS OF TRANSPORT COSTS COMBINED SAMPLE DATA

Dependent Variable	X_1*	X_2*	X_3*	X_4*	Intercept	r	Sample Size
Q_A	-3.011 (-.495)	-.468 (-.048)	-54.531 (-1.374)	68.886 (2.227)	84.124	.2766	35
Q	3.139 (1.214)	-10.883 (-2.604)	76.367 (3.850)	68.563 (3.845)	-110.297	.2455	140
M_A	-.006 (-1.609)	.004 (.691)	-.053 (-2.298)	-.017 (-.920)	.283	.2286	35
M	-.002 (-.694)	-.006 (-1.699)	.108 (6.497)	-.016 (-1.069)	.095	.3145	140

*Variables:
Q_A: Actual output
Q : Program generated output
M_A: Actual market share
M : Program generated market share
X_1*: Coal transport costs per ton of steel
X_2*: Ore transport costs per ton of steel
X_3*: Costs of transporting steel to nearest market
X_4*: Dummy variable for time

Table 5.7

SIGNIFICANCE AT .05 CONFIDENCE LEVEL OF THE DIFFERENCE BETWEEN COEFFICIENTS OF ACTUAL OUTPUT REGRESSIONS AND OPTIMAL OUTPUT REGRESSIONS-COMBINED SAMPLE (VALUES OF t STATISTIC IN PARENTHESES)

Dependent Variable	Independent Variables			
	X_1*	X_2*	X_3*	X_4*
Quantity of output	no (1.011)	no (1.065)	yes (3.298)	no (.010)
Market share	no (1.130)	yes (1.757)	yes (6.967)	no (.056)

Variables: X_1*: Coal transport cost
X_2*: Ore transport cost
X_3*: Transport cost to nearest market
X_4*: Dummy variable for time

coefficients of transport cost variables again are rather puzzling. They do indicate a predominantly negative relationship between coal transport costs and the output variable, and also between ore transport costs and the output variable, as would be expected. The coefficients of the transport cost to the nearest market are positive for program generated output and negative for actual output. The dummy variable has a positive coefficient when output level is used as the dependent variable and negative coefficient when market share is the dependent variable.

For this combined cross-sectional-time series sample, both regressions relating actual output variables to transport costs generate F values that are not significant at the .05 confidence level. The F values for the regressions relating program generated output to transport costs are both significant at the .05 confidence level.

Table 5.7, indicating the significance of the difference between the coefficients of actual and program generated output regressions for each independent variable when the combined sample is used, shows no significant difference for coal transport costs. The coefficients of ore transport costs differ significantly when market share is the dependent variable, but not when quantity of output is used as the dependent variable. The coefficients of transport cost to the nearest market differ significantly when either market share or output quantity is the dependent variable.

Although a definitive answer to the question of whether program generated and actual output levels differed significantly is not provided by the regression analysis, the analysis suggests the two output patterns did differ. The significance of the F values for the two sets of regressions supports this contention, as does the t tests on the regression coefficients. Further, the differing results for different years from the t tests on the coefficients in the actual output regressions also indicate a convergence of the actual and program generated output patterns over time.

III. DESCRIPTIVE ANALYSIS OF THE DIVERGENCE BETWEEN OPTIMAL AND ACTUAL LOCATIONS

At first glance, the locations of steel production generated by the linear programming model (heavily oriented towards Birmingham for the early years of the study) are surprising. Given the apparent divergence between actual and program-generated locations, some possible explanations include:

(1) the divergence is not as significant as first appears to be the case

(2) the industry for a variety of reasons did not locate so as to minimize transport costs

(3) the linear programming model used did not accurately depict all relevant transport cost related variables

(4) the data used were not sufficiently accurate to obtain realistic results.

Several points related to the above possibilities need to be considered. Although the program generated locations and actual locations were very different for the early years of the study, in both cases production moves heavily towards the Great Lakes area. The program therefore predicts a locational change similar to that which was occurring by the end of the era being studied.

Further, the program utilized data acquired *ex post facto*. If steel producers in 1879 (and/or their customers) believed Birmingham ores produced inferior steel, this perceived inferiority would have influenced locational decisions. Indeed, the Birmingham ores were generally inadequate for producing Bessemer steel; in 1879, open-hearth steel production was only beginning to overtake Bessemer steel tonnage. Steel producers in that early year probably felt uneasy about the quality of Birmingham ores; production at Birmingham in 1879 and 1889 using the more acceptable Great Lakes areas would have negated Birmingham's transport cost advantage. Thus, the shifting technological base may have given an advantage to Birmingham production which was not instantly perceived by actual and potential steel producers located elsewhere. If such an advantage existed, however, the Birmingham producers should have grown relative to those located elsewhere, thereby gradually relocating the

the industry. This did occur to a limited extent; Birmingham's share of actual steel production did increase but not dramatically over the forty years covered in the study.

The imperfect flow of information might have impeded the development of the Birmingham steel industry by affecting the flow of investment funds to that area. Thus, Birmingham's share of total production may have been limited by the insufficiency of money capital in that region during the late nineteenth century. Lance Davis, in studying the development of American financial intermediaries, indicated the uneven growth of lending institutions in different regions. He also outlined the unwillingness of northeastern savers to lend for unfamiliar projects outside the northeast. The immobility of funds was evidenced by the differences in interest rates among regions; northeastern rates were lower than southern interest rates throughout the era, although the gap was shrinking over time.[1] For steel producers, access

---

[1]Lance Davis, "The Investment Market; 1870-1914: The Evolution of a National Market", *Journal of Economic History*, Vol. XXV, No. 3 (September, 1965), pp. 355-399.

to money capital was an important consideration by the early 1880's because the average size of an efficient integrated steelworks, and hence the necessary initial investment, was substantial and growing in the late nineteenth century. Immobility of money capital may have limited the development of Birmingham as a steel producer until such immobility was overcome by U. S. Steel Corporation's interest in that location in the early twentieth century.

The program generated results imply considerable movement in the location of steel production, more movement than might be warranted, given the extensive amounts of capital equipment used by the industry. Might the location model be inadequate because it lacks any capacity constraint? Before adding an additional immobile factor of production constraint to the model, the following should be kept in mind:

(1) a decade elapses between each of the years involved in the study;

(2) a marked transformation in technology used, the shift from the Bessemer process to the open-hearth technique, occurred in the early years of the

study;

(3) demand for American steel grew rapidly over the forty years encompassed by the study.

The _Report of the Commissioner of Corporations on the Steel Industry_, in its inquiry into investments and profits in that industry, suggests that ". . . most manufacturing companies aim to make sufficient depreciation allowances, etc., to replace the capital invested in plant in about 20 years."[1] Without any consideration of technological or locational obsolesence, then, capital was rebuilt at about twenty year intervals. This estimate of lifetime was made for facilities in use between 1902 and 1906; there is no reason to assume dramatically different figures in earlier or later years. In addition to the twenty year life estimate, the report mentioned, among the steel firms' costs, a rebuilding fund designed to handle periodic rebuilding and relining of open-hearth furnaces. Such relining and rebuilding was necessary every few months to every few years

[1] _Report of the Commissioner of Corporations on the Steel Industry; Part III: Costs of Production_, Government Printing Office, 1913), p. 509.

depending upon the intensity of furnace use.[1] This fund, according to the figures reported by the Commissioner of Corporations, varied in size from thirty percent of the separate depreciation fund to one hundred percent of that fund per ton of open-hearth ingots produced.[2]

Physical obsolesence required replacement of over half the capital stock every ten years (50% to handle depreciation itself; a smaller to equal amount to cover rebuilding of furnaces). Technological obsolesence probably hastened the replacement of marginal facilities. Further consideration should also be given the rapid growth of output. With growing demand and production, locational change can occur without abandoning existing productive capital. This could occur by maintaining existing capital and locating new facilities in new areas accessible to growing demand; the relative importance of old production centers would gradually diminish as a result.

The linear programming model predicts the output levels given in Table 5.8. Reduction of

---

[1]Ibid, pp. 98, 151, 156.

[2]Ibid.

Table 5.8

PROGRAM GENERATED LEVELS OF OUTPUT BY LOCATION*

Location	Year				
	1879	1889	1899	1909	1919
Baltimore				7,139,571	
Birmingham	2,642,950	4,388,250	7,894,901	7,849,540	2,518,610
Buffalo					12,952,743
Cleveland			1,197,026	2,014,610	5,528,803
Pittsburgh		1,007,212			
Pueblo					

*The figures in this table do not reflect higher ore input requirements for Birmingham than for other production sites.

output for a particular location might have required abandonment of some capital equipment in that location. The economic desirability of such capital abandonment depends upon the transport cost saving as a result of optimal location as compared with the cost of abandoning such capital. An examination of these two components of the net saving from optimal location follows in the next two paragraphs. Figures in Table 5.8 give tonnages of output at the feasible production sites; to calculate the costs of abandoning capital, these must be converted to capital costs. To do so necessitates information on dollar values of capital required per ton of steel output. Data on capital requirements for nineteenth century production are neither readily available nor totally accurate. However, rough estimates of the dollar value of capital required per ton of steel output can be made. _The Census of Manufacturing_ gives capital costs for each of the years in the study. In some years, the costs are broken down into land, buildings and equipment, and working capital components; for other years, only a total dollar capital figure is given. To obtain the dollar cost of capital abandonment, only the costs of buildings (plant) and equipment should be considered. For those years for which the components of capital are given, plant and equipment amounts to about 45% of total capital. Therefore, 45% of the total capital is assumed to consist of buildings and equipment in those years for which a breakdown is not given. The dollar capital figure in the various censuses is usually a total value for steelworks and rolling mills. This dollar value for a particular year is therefore divided by total output tonnage of steelworks and rolling mills for that year to obtain a dollar cost of buildings and equipment per ton of output. These average (per ton) dollar capital requirements are shown in Table 5.9. The figure for 1919 is out of line with the other figures, but may represent price inflation during the World War I era; it may, alternatively, indicate diminishing returns to investment in steel producing capacity due to the industry's expanding beyond the infant industry size.

Table 5.10 indicates the required dollar capital for producing the levels of output generated by the program for the different locations; it is derived by multiplying tonnages of output (Table 5.8) by the dollar capital costs per ton (Table

Table 5.9

CAPITAL COSTS PER TON OF STEEL OUTPUT

Year	Costs
1879	14.30
1889	15.25
1899	13.23
1909	16.91
1919	33.01

Table 5.10

DOLLAR CAPITAL REQUIRED TO PRODUCE
PROGRAM GENERATED LEVELS OF OUTPUT

Location	Year				
	1879	1889	1899	1909	1919
Baltimore				120,730,416	
Birmingham	37,794,185	66,920,813	105,640,240	132,735,721	83,139,316
Buffalo					427,570,046
Chicago					
Cleveland			15,836,654	34,067,055	182,505,787
Pittsburgh		15,359,983			
Pueblo					

5.9).

As stated earlier in the chapter, the <u>Report of the Commissioner of Corporations</u> estimated a twenty year lifetime for steelworks and rolling mills, if adequately maintained; this lifetime could be shortened by forgoing relining of the furnace and related repairs. The relining process was fairly expensive, as indicated by the substantial sums set aside annually for this purpose (sometimes as large as the depreciation account). Therefore, the capital available at a particular site for production at time t is assumed to be 40% of that available at time t-1 (additions can be made to this capacity; we are concerned here primarily with the amount of capacity requiring abandonment if program generated optimal locations were pursued). The remaining 60% is a rough guess of the amount of capacity no longer useful due to physical obsolesence and technological obsolesence, and is probably a conservative calculation (underestimate) of capacity depletion over a decade. Table 5.11 indicates the excess (carryover) capacity at each location and date for which there is an excess. It results from multiplying capacity requirements for a particular date by .4 (100%-

60%) and subtracting the result from the capacity requirements of the next decade.

The excess capacities of Table 5.11 represent a cost of optimal location. These costs should be compared with the benefits of optimal (minimum transport cost) locations. The benefits are measured in terms of the reductions in transportation costs. To obtain these reductions, transport costs with optimal location, not constrained by capacity considerations, are compared with the transport costs when output at any site is constrained not to exceed the tonnage actually produced at that site. The transport costs resulting from solution of each of these linear programming problems are summarized in Table 5.12. (These represent the value of the minimized objective function for the linear program when (a) a capacity constraint is included in the linear program and (b) no such capacity constraint is included).

An immediate reaction to these costs involves their large size; given these figures, transport costs accounted for a substantial portion of the total price of steel. This observation fits the comments found in contemporary descriptions of

Table 5.11

EXCESS CAPACITY (DOLLAR VALUE)

Location	Year				
	1879	1889	1899	1909	1919
Baltimore					$48,292,058
Birmingham					
Buffalo					
Chicago					
Cleveland					
Pittsburgh			$6,143,993		
Pueblo					

Table 5.12

TRANSPORT COSTS: NET SAVING AS A RESULT OF OPTIMAL LOCATION

	1879	1889	1899	1909	1919
(a) Transport Costs Capacity Constrained Program	67,678,800	49,841,632	89,213,552	178,593,710	214,831,010
(b) Transport Costs No Capacity Constraints	33,230,016	36,219,120	71,035,920	125,952,670	192,477,070
(c) Difference (a) (b)	34,448,784	13,622,512	18,177,632	52,641,040	22,353,940

the steel industry in the United States.[1] Nonetheless, the figures are so large, especially in the earlier years of the study, that they suggest somewhat inflated transport rate estimates. With the difficulty of obtaining nineteenth century freight rate data and the widespread practice of allowing rebates on rail rates to preferred (usually large) customers, the transport rate data used in the linear programs may be biased upward. If such a bias exists, and if established steel producers did receive rate rebates, minimum transport cost locations would have more likely been the established centers (such as Pittsburgh) than the newer sites (such as Birmingham). Freight rebates became less likely with the establishment in 1887 and the political development thereafter of the Interstate Commerce Commission; and, indeed, the gap between program generated optimal output and actual output at various sites does diminish in the later years of the study.

Returning to the benefits and costs of

---

[1]For example, James Swank commented on the large transport costs paid by U. S. iron and steel producers relative to European transport costs in his statement on the industry accompanying the industry data in the Census of 1880.

optimal location as determined by the linear programming model, these benefits and costs occur at varying points in time. To compare them appropriately, they need to be brought to a single focal date. Thus, the values of transport cost savings and abandoned capacity costs are discounted to obtain the present discounted value of each in 1879. A discount rate of 5% per annum was used. To compare costs and benefits, therefore, we compare

(5.4) PV (Costs of abandoning capital)

$$= \frac{6,143,993}{(1.05)^{20}} + \frac{48,292,058}{(1.05)^{40}}$$

$$= 9,175,297$$

with

(5.5) PV (Transport cost savings)

$$=34,448,784+\frac{13,622,512}{(1.05)^{10}}+\frac{18,177,632}{(1.05)^{20}}+\frac{52,641,040}{(1.05)^{30}}$$

$$+\frac{22,353,940}{(1.05)^{40}}$$

$$=65,017,985.$$

These results imply that the costs of abandoning some capital did not outweigh the substantial monetary benefits, through reduced transport costs, of optimal location. Further, even if the entire capacity of 1879 were to be relocated, costs of abandoned capital would not increase sufficiently to warrant dismissing the minimum transport cost locations.

The linear programming model also ignores scrap as a possible substitute for iron ore. In the twentieth century, scrap steel becomes an increasingly important component in new steel production. The possibility of using steel scrap in production would enhance the attractiveness of the major steel demand centers for previous years in locating new steel production facilities. However, in the time period studied, scrap usage is not extremely important. Further, most of the scrap used at this time was generated internally by the steelworks which used it. For later years, the model would have to be adjusted to include possible inputs of scrap; to do so, input requirements would have to include the possibility of substituting scrap for some iron ore and the available supplies of scrap (dependent upon demand for steel in earlier years) would enter the problem as constraints. The transport costs for scrap shipments would enter an adjusted objective function.

A final consideration about the validity of the program generated locations of the steel industry concerns the data used in the study.

Extensive amounts of time and effort were involved in obtaining the data. However, the nineteenth century published data are not always accurate; additionally, some data components were not directly available, and were constructed. Therefore, the data used must be considered hard work estimates and not absolute truths concerning the size of the variables involved.

In summary, the results of the program are not as surprising as they first appear; leaving a capacity constraint or cost variable out of the linear programming model is probably not damaging and the same is likely true for the exclusion of scrap between 1879 and 1919; and the data used must be assumed to be only estimates of the true values of the variables in the study.

IV. CONCLUSIONS

The linear programming model used with historical data to generate minimum transport cost locations of the steel industry has unquestionably provided some intriguing results. Whether these results imply an industry which located in Weberian minimum transport cost fashion between 1879 and 1919 is more difficult to ascertain. The industry actually located predominantly in the Pittsburgh area, whereas the model suggests that Birmingham would have been the minimum transport cost location in 1879. Over the following forty years, changes occurred in demand, in transport rates, and in the tonnages of coal and ore required per ton of steel. These changes influenced both the actual and program generated levels of steel output at the various production sites. Both actual and program generated output locations moved northwestward to some extent when the entire time period is considered, although program generated locations did not move westward as strongly as did actual output. These northwestward movements occurred despite the eastward movement of demand, perhaps suggesting that the locational changes in the industry were not dominated by demand considerations.

The actual production of steel moved further westward than did program generated output, with Cleveland (Ohio) and Chicago (Indiana-Illinois) becoming relatively more important producers over time. The program results grant Buffalo and Cleveland the leading positions in this movement (and Buffalo is not west of Pittsburgh except in its transportation links with western resource suppliers and markets). Buffalo, Cleveland, and Chicago-Gary are closer to each other from a transport cost perspective than they are in actual mileage, because all are located on the Great Lakes. By the latter half of the nineteenth century, cheap transportation routes were available connecting the various lakes. Therefore, the movement towards Chicago and Cleveland in actual production is not dramatically different from the movement of program generated production towards Cleveland and Buffalo. The statistical tests used to examine the differences between actual and optimal output did not take into account the distances between various sites. Thus, in the tests, differences between producing in Buffalo and in Cleveland are counted as heavily as differences between Birmingham and Chicago. With more variables in a regression analysis, part of this problem could be eliminated. However, the limited number of observations per sample on actual output prevent expanding the regression analysis to include more independent variables.

Careful comparision of the locational pattern predicted by the linear program with that which actually occurred, when combined with qualitative and quantitative information on the industry and the time period, suggests the optimal and actual locations of the steel industry were very different in 1879. Similar consideration of changes in optimal and actual output patterns suggests the two were moving closer together over time. To further buttress the conclusion that the steel industry became more sensitive to transport costs in its locational choices between 1879 and 1919, a comparision of actual transport costs (given the actual locations of steel production) with transport costs when production occurs at minimum transport cost sites is pertinent. If actual transport costs came closer over time to the minimum feasible transport costs, it is likely that actual locations were moving closer to minimum transport cost locations.

An estimate of actual transport costs for a given year is provided by the linear programming model when output at each production site is constrained not to exceed actual production at that site. This estimate, the value of the objective (transport cost) function for the capacity constrained program, is indicated for 1879, 1889, 1899, 1909, and 1919 in Table 5.12; the same table also gives the value of the objective function when output at each location is not constrained. Table 5.13 indicates the difference between the two as a fraction of transport costs when output at each site is constrained (the estimate of actual transport costs). These figures therefore are an estimate of the fraction of transport costs which could have been saved via optimal location. The fraction declines for each ten year period except that between 1899 and 1909; the overall decline over the forty year period is substantial. If this fraction is regressed on minimum transport costs (divided by 10,000,000 to ease computational difficulties), the resulting regression has a coefficient of determination equalling .50

Table 5.13

TRANSPORT COST REDUCTION DUE TO OPTIMAL LOCATION AS A FRACTION OF ESTIMATED ACTUAL TRANSPORT COSTS

Year	Transport Cost Reduction Divided by Actual Transport Costs Estimate
1879	.509
1889	.273
1899	.204
1909	.295
1919	.104

(correlation coefficient of .70). The coefficient of the independent variable is -.02, with a standard error of .01. Given a sample size of five, this coefficient significantly differs from zero at the .10 confidence level. These results reenforce earlier suggestions of a movement by the steel industry towards minimum transport cost locations.

Thus, considerable evidence points to differences between actual and optimal locations in the early years of the study, but with the differences becoming smaller over time. The evidence from the three statistical tests is less clearcut but suggests the same general conclusions. All three tests suffer from the limited number of observations on actual output in any one year. For both nonparametric tests, the differences between optimal and actual output become significant when pooled (and larger sample) data are used. The limitations of the regression analysis for individual cross-sectional samples has already been discussed. It seems likely that with an expanded sample size the statistical tests would provide additional support for the conclusion that (1) the steel industry did not locate initially so as to minimize transport costs; and (2) the steel industry between 1879 and 1919 was moving toward minimum transport cost locations.

Why such a movement would have occurred is an intriguing question. During the later years of the period studied, the basing point pricing system was introduced. The delivered price of steel was quoted as the price in Pittsburgh plus transport costs from Pittsburgh to the demand site, regardless of the point of origin.[1] This pricing mechanism at least suggests that Pittsburgh producers were not sanguine about that region's comparative advantage in steel production. However, even with Pittsburgh plus pricing, profits with minimum transport cost location (ceteris paribus) would exceed those profits without minimum transport cost location. Although some nonoptimal locations are allowed a longer survival than they would have in the absence of such a price fixing scheme, there is still a profit incentive for efficient location. And the scheme was introduced at a time when location was moving towards minimum transport

---

[1]Other basing points were added later, so that the single basing point system became a multiple basing point system.

cost sites.

It might be argued that transport costs were too small a fraction of total costs to affect locational decisions in the early years of the study. However, transport rates fell over the time interval being studied (although they were rising between 1909 and 1919, the rates had not reached the 1879 level by 1919). Transport costs were an important though declining component of total steel production costs, ranging from roughly 33% in 1879 to 10% around 1919. These percentages are too high, presumably, to have been ignored by a profit maximizing and mobile producer.

Perhaps actual locations approached the optimal more closely over time because producers were able to adjust to the new technology with which the period 1879-1919 began; in 1879, the average producer may have had insufficient information for choosing minimum transport cost locations. Although not extensively explored in this study, the greater industrial concentration of the industry in later years appears to have been associated with greater geographic dispersion and a locational pattern closer to the minimum transport cost one. Whether this was due to more centralized decision making and access to more information on the part of larger firm(s) with substantial market power cannot be answered here. The geographic dispersion may have occurred in the absence of increased industrial concentration with adjustments to new technology and freer flow of resources (including information) over time. On the other hand, the greater concentration may have facilitated a broader management view and greater concern for efficiency in locational decisions.

In summary, this study of the location of the steel industry between 1879 and 1919 provides some interesting observations on that industry's locational patterns in given years and over time. Additionally, it raises a number of intriguing questions about locational decisions. For example, was otherwise efficient industrial development in the postbellum South retarded by capital immobility? did industrial structure influence, favorably or otherwise, industrial location? were transport costs (especially those subject to regulation) sufficiently important in business decisions to relocate substantial amounts of industrial activity? Thus, the linear programming model of industrial location, applied in a historical context, allows not only some interesting, if tentative, conclusions about the location of the steel industry, it also raises some fascinating additional questions. These attributes suggest it is a useful tool in describing and analyzing the historical pattern of industrial location.

APPENDIX A: DATA USED IN THE SOLUTION OF THE MODEL

I. The Problem

Solution of the linear programming model requires data on the fixed demand at every consumption site, output per ton of iron ore, output per ton of coal, and transport costs per ton shipped for all possible ore, coal, and steel shipments. The derivations of the first three items are explained in the body of the paper, primarily in Chapter three. The fourth item, transportation cost data, will be examined in greater detail in this Appendix, as will the output-input coefficients for Alabama. Actual solution of the model was accomplished by using an IBM 370 computer with the linear programming program, Simplex.

II. Transport Cost Data

For each year of the study, 91 possible shipments appear in the program's objective function, i.e., the transportation cost function to be minimized. The appropriate transport costs appear as coefficients of these shipments. With five years in the study, a total of 455 transport costs between 1879 and 1919 must be obtained. Not too surprisingly, all 455 costs are not readily available. The general procedure followed here is to locate those transport costs that are reasonably available; a remarkable amount of transport cost information is available on this industry because of the interest in iron and steel production. Once a rate is established between two points in a given area, the corresponding ton-mile rate is calculated for that region and category of freight. That rate is then multiplied by the transport mileage between the two points for which the transport cost is being sought. By this mechanism, most gaps in the data can be filled. On deliveries of foreign ore to Baltimore, freight rates can be obtained for some, but not all, years. In this case, North's index of ocean freight rates is used to obtain rates in years for which there is no direct observation.[1]

For 1879, the major source of data is Swank's report on the United States iron and steel

---

[1]Douglass C. North, "Ocean Freight Rates and Economic Development," *Journal of Economic History*, XVIII (December, 1958), pp. 537-555.

industry, prepared for the 1880 census of manufactures.[1] A second source is the Census of Transportation for 1880. Swank gives rates of the shipments of ore, coal, and steel between major sites. Only very limited reference is made to southern transport rates, but all references indicate southern rates were lower than northern freights. Unless there is a direct observation for the southern rate, the northern ton-mile rate is used. In all cases, the cost of moving raw material from a source adjacent to production is based on the cost of shipping coal from Connellsville to Pittsburgh; again, this is likely to overstate southern transport costs. In every year, the transport cost of foreign ore is the ocean freight to Baltimore plus the overland freight from Baltimore to the given production site.

For 1889, the Commissioner of Labor's study on the costs of producing iron, steel, coal, etc., provides extensive data on rail and water freight rates for ore, coal, and steel between a variety

---

[1]United States, Department of the Interior, Census Office, Report on the Manufactures of the United States at the Tenth Census, 1880, (Washington, D. C.: Government Printing Office, 1883).

of points.[1] The report by the National Waterways Commission gives Great Lakes freight rates. Henry Fink, in his volume on railway rates, gives ton-mile freights for selected years, products, and routes.[2] Noyes has rates on classified traffic for a number of years up to 1900 on railroads operated between New York and Chicago, the Pennsylvania Railroad, and several midwestern roads.[3] Warren in his recent book on the American steel industry quotes several transport rates beginning with the late 1880's.[4]

For both the turn of the century and 1910, the Report of the Commissioner of Corporations yields much data on transport rates, and is the

---

[1]United States, Department of the Interior, Census Office, Report on the Manufacturing Industries in the United States at the Eleventh Census, 1890, Part III, Selected Industries, (Washington, D. C.: Government Printing Office, 1895).

[2]Henry Fink, Regulation of Railway Rates on Interstate Freight Traffic, (New York: The Evening Post Job Printing Office, 1905).

[3]Walter C. Noyes, American Railroad Rates, (Boston: Little Brown and Company, 1906).

[4]Kenneth Warren, The American Steel Industry, 1850-1970: A Geographical Interpretation, (Oxford: Clarendon Press, 1973.)

major source of information.[1] Also, the Digest of Hearings on Railway Rates, in comparing American rail rates with those in Germany on a variety of products, provides much information on turn of the century rail rates.[2] The same source gives data on rail rates issued to meet water route competition. An ICC report, A Forty Year Review of Changes in Freight Tariffs, published in 1903 as a separate appendix of its 1902 annual report, contains numerous rail transport rates, especially for 1889 and 1899.[3]

The major data source for 1919 is the report of the U. S. Tariff Commission on preferential tariff rates in the U. S.; this source contains large amounts of data on rates for imported and exported items and the differentials

---

[1]United States, Bureau of Corporations, Report of the Commissioner of Corporations on the Steel Industry, (Washington, D. C.: Government Printing Office 1913).

[2]United States, Congress, Regulation of Railway Rates, Digest of the Hearings, Document 244, 59th. Congress, 1st. Session, 1905.

[3]United States, Interstate Commerce Commission, Annual Report on the Railways in the United States in 1902, Pt. II, A Forty Year Review of Changes in Freight Rates, (Washington, D. C. Government Printing Office, 1903).

between these and domestic rates.[1]

III. Input-Output Coefficients for Alabama

The linear programming model actually uses output per ton of ore input ($1/\alpha_r$) and output per ton of coal input ($1/\alpha_c$), obtained by calculating the reciprocal of the input-output ratios. For the national figures, the results are given in Chapter 3. The calculations are more difficult and also more tenuous for Alabama separately because no data is reported separately by the census on steelworks and rolling mills in that state for most years. Therefore, ore and coal requirements per ton of pig iron are calculated for Alabama. Then the national figure on additional coal and ore used to convert pig iron to steel is applied to Alabama's pig iron figures to obtain the inputs per ton of steel output. The resulting figures are rather erratic, but definitely greater than the national average; Alabama therefore has lower outputs per ton of coal and ore inputs than the rest of the country. Figures on α_r and α_c are given

---

[1]United States, Tariff Commission, *Preferential Transportation Rates and Their Relation to Import and Export Traffic of the United States*, (Washington, D. C.: Government Printing Office, 1922).

for Alabama in Table A-1 below.

Table A-1

INPUT-OUTPUT COEFFICIENTS FOR ALABAMA

	α_e	α_r
1879	3.82	2.54
1889	5.92	8.20
1899	3.15	1.93
1909	2.43	3.95
1919	1.36	2.38

APPENDIX B: SHIPMENTS OF COAL, ORE IN THE CAPACITY CONSTRAINED MODEL

The levels of output at various production sites implied by the constrained capacity model are given in Chapter 4. Table B-1, below, indicates the ore and coal shipments necessary to provide this production.

Table B-1

RAW MATERIALS SHIPMENTS UNDER THE CONSTRAINED CAPACITY MODEL

	1879	
Coal shipments:	Connellsville to Baltimore	2,175,744
	Birmingham to Birmingham	1,994
	Connellsville to Buffalo	769,028
	Connellsville to Chicago	1,227,757
	Connellsville to Cleveland	1,168,923
	Connellsville to Pittsburgh	2,763,760
Ore shipments:	Foreign to Baltimore	1,496,397
	Birmingham to Birmingham	1,371
	Gt. Lakes to Buffalo	528,909
	Gt. Lakes to Chicago	844,407
	Gt. Lakes to Cleveland	803,943
	Gt. Lakes to Pittsburgh	1,900,813

Table B-1 Continued

RAW MATERIALS SHIPMENTS UNDER THE CONSTRAINED CAPACITY MODEL

	1889	
Coal shipments:	Birmingham to Birmingham	102,363
	Connellsville to Buffalo	470,639
	Connellsville to Chicago	923,757
	Connellsville to Cleveland	2,211,789
	Connellsville to Pittsburgh	6,870,786
Ore shipments:	Birmingham to Birmingham	82,997
	Gt. Lakes to Buffalo	381,599
	Gt. Lakes to Chicago	748,992
	Gt. Lakes to Cleveland	1,793,342
	Gt. Lakes to Pittsburgh	5,570,907
	1899	
Coal shipments:	Connellsville to Baltimore	4,004,317

Table B-1 Continued

RAW MATERIALS SHIPMENTS UNDER THE CONSTRAINED CAPACITY MODEL

	Birmingham to Birmingham	209,919
	Connellsville to Buffalo	267,405
	Connellsville to Chicago	3,704,054
	Connellsville to Cleveland	5,305,226
	Connellsville to Pittsburgh	4,303,509
Ore shipments:	Foreign to Baltimore	3,538,059
	Birmingham to Birmingham	185,476
	Gt. Lakes to Buffalo	236,269
	Gt. Lakes to Chicago	3,272,758
	Gt. Lakes to Cleveland	4,687,492
	Gt. Lakes to Pittsburgh	3,802,414
	1909	
Coal shipments:	Connellsville to Baltimore	4,409,795

Table B-1 Continued

RAW MATERIALS SHIPMENTS UNDER THE CONSTRAINED CAPACITY MODEL

	Birmingham to Birmingham	544,725
	Connellsville to Buffalo	1,451,317
	Connellsville to Chicago	5,547,804
	Connellsville to Cleveland	5,631,681
	Connellsville to Pittsburgh	13,330,572
Ore shipments:	Foreign to Baltimore	7,196,999
	Birmingham to Birmingham	889,018
	Gt. Lakes to Buffalo	2,386,617
	Gt. Lakes to Chicago	9,054,283
	Gt. Lakes to Cleveland	9,191,173
	Gt. Lakes to Pittsburgh	21,756,112

Table B-1 Continued

RAW MATERIALS SHIPMENTS UNDER THE CONSTRAINED CAPACITY MODEL

	1919	
Coal shipments:	Connellsville to Baltimore	2,270,961
	Birmingham to Birmingham	900,796
	Connellsville to Buffalo	1,224,554
	Connellsville to Chicago	3,103,562
	Connellsville to Cleveland	6,515,178
	Connellsville to Pittsburgh	13,436,164
Ore shipments:	Foreign to Baltimore	3,102,296
	Birmingham to Birmingham	1,230,551
	Gt. Lakes to Buffalo	1,672,829
	Gt. Lakes to Chicago	4,239,689
	Gt. Lakes to Cleveland	8,900,202
	Gt. Lakes to Pittsburgh	18,354,752

REFERENCES

Journal Articles

Davis, Lance E. "The Investment Market, 1870-1914: The Evolution of a National Market." Journal of Economic History, XXV (September, 1965), 355-399.

Harris, Chauncy D. "The Market as a Factor in the Localization of Industry." Annals of the Association of American Geographers, XLIV (December, 1954), 315-348.

Isard, Walter. "Some Locational Factors in the Iron and Steel Industry Since the Early Nineteenth Century." Journal of Political Economy, LVI (June, 1948), 203-217.

-------, and Capron, William M. "The Future Locational Pattern of Iron and Steel Production in the United States." Journal of Political Economy, LVII (April, 1949), 118-133.

Hotelling, Harold. "Stability in Competition." The Economic Journal, XXXIX (1929), 41-57.

Mills, Edwin and Lav, Michael. "A Model of Market Areas with Free Entry." The Journal of Political Economy, LXII (June, 1964), 278-288.

North, Douglass C. "Ocean Freight Rates and Economic Development." Journal of Economic History, XVIII (December, 1958), 537-555.

Smithies, Arthur. "Optimum Location in Spatial Competition." Journal of Political Economy, XLIX (June, 1941), 423-439.

Temin, Peter. "The Composition of Iron and Steel Products, 1869-1909." Journal of Economic History, XXIII (December, 1963), 447-471.

Walters, Alan A. "Production and Cost Functions: An Econometric Survey." Econometrica, XXXI (January-April, 1963), 1-66.

Books

Adams, Walter. "The Steel Industry." The Structure of American Industry. Edited by Walter Adams. New York: The Macmillan Company, 1971.

Alderfer, E. B. and Michl, H. E. Economics of American Industry. New York: McGraw-Hill Book Company, Inc., 1942.

Andreano, Ralph, ed. New Views on American Economic Development. Cambridge, Massachusetts: Schenkman Publishing Company, 1965.

Beckmann, Martin. Location Theory. New York: Random House, 1968.

Clark, Victor S. History of Manufactures in the United States, Vol. I, II, III. New York: Peter Smith, 1949.

Danø. Sven. Linear Programming in Industry. New York: Springer-Verlag, 1974.

Dantzig, George B. "Maximization of a Linear Function of Variables Subject to Linear Inequalities." Activity Analysis of Production and Allocation. Edited by Tjalling Koopmans. John Wiley and Sons, Inc., 1951.

Daugherty, Carroll R.; deChazeau, Melvin G.; Stratton, Samuel S. The Economics of the Iron and Steel Industry, Vol. I. New York: McGraw Hill Book Company, Inc., 1937.

Derry, T. K. and Williams, Trevor I. A Short History of Technology. London: Oxford University Press, 1960.

Dorfman, Robert; Samuelson, Paul A.; and Solow, Robert M. Linear Programming and Economic Analysis. New York: McGraw-Hill Book Company, Inc., 1958.

Fink, Henry. Regulation of Railway Rates on Interstate Freight Traffic. New York: The Evening Post Job Printing Office, 1905.

Fisher, Douglas Alan. The Epic of Steel. New York: Harper & Row, Publishers, 1963.

Freund, John. Modern Elementary Statistics. 3rd. ed. Englewood Cliffs, New Jersey: Prentice-Hall, Inc., 1967.

Henderson, James M. and Quandt, Richard E. Microeconomic Theory: A Mathematical Approach. New York: McGraw-Hill Book Company, 1958.

Hoover, Edgar. The Location of Economic Activity. New York: McGraw Hill Book Company, Inc., 1948.

--------. Location Theory and the Shoe and Leather Industries. Cambridge, Massachusetts: Harvard University Press, 1937.

Hughes, Jonathan. The Vital Few: American Economic Progress and Its Protagonists. London: Oxford University Press, 1965.

Hunter, Louis. "The Heavy Industries." Growth of the American Economy. Edited by Harold F. Williamson. Englewood Cliffs, New Jersey: Prentice Hall, Inc., 1951.

Isard, Walter. Location and Space Economy: A General Theory Relating to Industrial Location, Market Areas, Land Use, Trade and Urban Structure. Cambridge, Massachusetts: The M.I.T. Press, 1956.

Kirkland, Edward. "Building American Cities." Views of American Economic Growth: The Industrial Era. Edited by Thomas C. Cochran and Thomas B. Brewer. New York: McGraw Hill Book Company, 1966.

Kroos, Hermann E. and Gilbert, Charles. American Business History. Englewood Cliffs, New Jersey: Prentice-Hall, Inc., 1972.

Lefeber, Louis. Allocation in Space: Production, Transport, and Industrial Location. Amsterdam: North-Holland Publishing Co., 1958.

Leontief, Wassily. The Structure of the American Economy, 1919-1929. New York: Oxford Press, 1951.

Lösch, August. The Economics of Location. Translated by William H. Woglom with the assistance of Wolfgang F. Stolper. New Haven: Yale University Press, 1954.

Mitchell, Wesley and Burns, Arthur. Measuring Business Cycles. New York: National Bureau of Economic Research, 1947.

Noyes, Walter Chadwick. American Railroad Rates. Boston: Little, Brown, and Company, 1906.

Nutter, G. Warren. Extent of Enterprise Monopoly in the United States, 1899-1939. Chicago: University of Chicago Press, 1951.

Paullin, Charles O. Atlas of the Historical Geography of the United States. Washington: Carnegie Institution of Washington, 1932. The volume was published jointly with the American Geographical Society of New York, and printed by A. Hoen & Co., Inc., Baltimore, Maryland.

Popplewell, Frank. Some Modern Conditions and Recent Developments in Iron and Steel Production in America. Manchester: University Press, 1906.

Siegel, Sidney. Nonparametric Statistics for the Behavioral Sciences. New York: McGraw-Hill Book Company, Inc. 1956.

Swank, James M. History of the Manufacture of Iron in All Ages, and Particularly in the United States from Colonial Times to 1891. 2nd. ed. Philadelphia: American Iron and Steel Association, 1892.

Temin, Peter. Iron and Steel in Nineteenth Century America: An Economic Inquiry. Cambridge, Massachusetts: M.I.T. Press, 1964.

Wall, Joseph Frazier. Andrew Carnegie. New York: Oxford University Press, 1970.

Warren, Kenneth. The American Steel Industry, 1850-1970: A Geographical Interpretation. Oxford: Clarendon Press, 1973.

Weber, Alfred. Theory of the Location of Industries. Translated, edited, and introduction by Carl J. Friedrich. Chicago: University of Chicago Press, 1929.

Government Documents

National Resources Planning Board. Industrial Location and National Resources. Washington: United States Government Printing Office, 1943.

U. S. Bureau of Corporations. Report of the Commissioner of Corporations on the Steel Industry, Part I. Organization, Investment, Profits, and Position of United States Steel Corporation. Washington, D. C.: Government Printing Office, 1911.

-------. Report of the Commissioner of Corporations on the Steel Industry. Part III. Cost of Production: Full Report. Washington, D. C.: Government Printing Office, 1913.

-------. Report of the Commissioner of Corporations on Transportation by Water in the United States, Pt. III. Washington: Government Printing Office, 1910.

U. S., Commissioner of Labor, 6th Annual Report, 1890. Cost of Production: Iron, Steel, Coal, etc. Washington, D. C.: Government Printing Office, 1891.

U. S. Congress. Regulation of Railway Rates, Digest of the Hearings. Doc. 244, 59th. Cong., 1st. sess., 1905.

U. S. Congress. House. Report on the Agencies of Transportation in the United States. 47th Cong., 2nd sess., 1882-1883. The Miscellaneous Documents of the House of Representatives, Vol. XIII, pt. 4.

-------. Report on the Transportation Business in the United States at the Eleventh Census: 1890: pt. I. Transportation by Land. 52nd. Cong., 1st. sess., 1891-92. The Miscellaneous Documents of the House of Representatives, vol. L, pt. II.

U. S. Congress, Senate. Final Report of the National Waterways Commission. S. Doc. 15, 62nd. Cong., 2nd sess., 1911-1912.

-------. Preliminary Report of the Inland Waterways Commission. S. Doc. 326, 60th. Cong., 1st. sess., 1908.

-------. Committee on Finance. Wholesale Prices, Wages, and Transportation. S. Rept. 1394, 52nd. Cong., 2nd sess., 1890.

U. S. Department of Agriculture. Division of Statistics. Changes in the Rates of Charge for Railway and Other Transportation Services, by H. T. Newcomb. Bulletin No. 15. Washington, D. C.: Government Printing Office, 1901.

U. S. Department of Commerce. Bureau of the Census. Historical Statistics of the United States: Colonial Times to 1957. Washington, D. C.: Government Printing Office, 1960.

-------. Bureau of the Census. Thirteenth Census of the United States Taken in the Year 1910, Vol. X, Manufactures: Reports for Principal Industries. Washington, D. C.: Government Printing Office, 1913.

-------. Bureau of the Census. Thirteenth Census of the United States Taken in the Year 1910, Vol. II, Population. Washington, D. C.: Government Printing Office, 1913.

-------. Bureau of the Census. Census of Manufactures, 1914, Vol. II, Reports for Selected Industries and Detail Statistics for Industries, by States. Washington, D. C.: Government Printing Office, 1919.

-------. Bureau of the Census. Fourteenth Census of the United States Taken in the Year 1920, Vol. X, Manufactures, Reports for Selected Industries. Washington, D. C.: Government Printing Office, 1923.

-------. Bureau of the Census. Fourteenth Census of the United States Taken in the Year 1920, Vol. III, Population. Washington, D.C.: Government Printing Office, 1922.

U. S. Department of Commerce and Labor. Bureau of the Census. Census of Manufactures, 1905. pt. IV; Special Reports on Selected Industries. Washington, D. C.: Government Printing Office, 1908.

U. S. Department of the Interior. Census Office. Report on the Manufactures of the United States at the Tenth Census, 1880. Washington, D. C.: Government Printing Office, 1883.

-------. Census Office. Report on Manufacturing Industries in the United States at the Eleventh Census: 1890. Part III, Selected Industries. Washington, D. C.: Government Printing Office, 1895.

-------. Census Office. Report on Manufacturing Industries in the United States at the Eleventh Census: 1890, pt. III. Selected Industries. Washington, D. C.: Government Printing Office, 1895.

-------. Census Office. Report on Population of the United States at the Eleventh Census: 1890, pt. I. Washington, D. C.: Government Printing Office, 1895.

-------. Census Office. Twelfth Census of the United States Taken in the Year 1900, Vol. X, Manufactures, pt. IV, Special Reports on Selected Industries. Washington, D. C.: United States Census Office, 1902.

-------. Census Office. Twelfth Census of the United States, Taken in the Year 1900. Vol. I, Population, pt. I, Washington, D. C.: United States Census Office, 1901.

U. S. Geological Survey. Mineral Resources of the United States. Washington, D. C.: Government Printing Office, 1890, 1900, 1910, 1920.

U. S. Interstate Commerce Commission. Annual Report on the Railways in the United States in 1902, pt. II, A Forty Year Review of Changes in Freight Rates. Washington, D. C.: Government Printing Office, 1903.

-------. Third Annual Report on the Statistics of Railways in the United States to the Interstate Commerce Commission for the Year Ending June 30, 1890. Washington, D. C.: Government Printing Office, 1891.

-------. Thirteenth Annual Report on the Statistics of Railways in the United States to the Interstate Commerce Commission. Washington, D. C.: Government Printing Office, 1901.

-------. Twenty-Third Annual Report on the Statistics on Railways in the United States to Interstate Commerce Commission. Washington, D.C.: Government Printing Office, 1911.

-------. Thirty-Third Annual Report on the Statistics of Railways in the United States to the Interstate Commerce Commission. Washington, D. C.: Government Printing Office, 1921.

U. S. Tariff Commission. *Preferential Transportation Rates and Their Relation to Import and Export Traffic of the United States.* Washington, D. C.: Government Printing Office, 1922.

Dissertations in American Economic History

An Arno Press Collection

1977 Publications

Ankli, Robert Eugene. **Gross Farm Revenue in Pre-Civil War Illinois.** (Doctoral Dissertation, University of Illinois, 1969). 1977

Asher, Ephraim. **Relative Productivity, Factor-Intensity and Technology in the Manufacturing Sectors of the U.S. and the U.K. During the Nineteenth Century.** (Doctoral Dissertation, University of Rochester, 1969). 1977

Campbell, Carl. **Economic Growth, Capital Gains, and Income Distribution:** 1897-1956. (Doctoral Dissertation, University of California at Berkeley, 1964). 1977

Cederberg, Herbert R. **An Economic Analysis of English Settlement in North America, 1583-1635.** (Doctoral Dissertation, University of California at Berkeley, 1968). 1977

Dente, Leonard A. **Veblen's Theory of Social Change.** (Doctoral Dissertation, New York University, 1974). 1977

Dickey, George Edward. **Money, Prices and Growth;** The American Experience, 1869-1896. (Doctoral Dissertation, Northwestern University, 1968). 1977

Douty, Christopher Morris. **The Economics of Localized Disasters:** The 1906 San Francisco Catastrophe. (Doctoral Dissertation, Stanford University, 1969). 1977

Harper, Ann K. **The Location of the United States Steel Industry, 1879-1919.** (Doctoral Dissertation, Johns Hopkins University, 1976). 1977

Holt, Charles Frank. **The Role of State Government in the Nineteenth-Century American Economy, 1820-1902:** A Quantitative Study. (Doctoral Dissertation, Purdue University, 1970). 1977

Katz, Harold. **The Decline of Competition in the Automobile Industry, 1920-1940.** (Doctoral Dissertation, Columbia University, 1970). 1977

Lee, Susan Previant. **The Westward Movement of the Cotton Economy, 1840-1860:** Perceived Interests and Economic Realities. (Doctoral Dissertation, Columbia University, 1975). 1977

Legler, John Baxter. **Regional Distribution of Federal Receipts and Expenditures in the Nineteenth Century:** A Quantitative Study. (Doctoral Dissertation, Purdue University, 1967). 1977

Lightner, David L. **Labor on the Illinois Central Railroad, 1852-1900:** The Evolution of an Industrial Environment. (Doctoral Dissertation, Cornell University, 1969). 1977

MacMurray, Robert R. **Technological Change in the American Cotton Spinning Industry, 1790 to 1836.** (Doctoral Dissertation, University of Pennsylvania, 1970). 1977

Netschert, Bruce Carlton. **The Mineral Foreign Trade of the United States in the Twentieth Century:** A Study in Mineral Economics. (Doctoral Dissertation, Cornell University, 1949). 1977

Otenasek, Mildred. **Alexander Hamilton's Financial Policies.** (Doctoral Dissertation, Johns Hopkins University, 1939). 1977

Parks, Robert James. **European Origins of the Economic Ideas of Alexander Hamilton.** (M. A. Thesis, Michigan State University, 1963). 1977

Parsons, Burke Adrian. **British Trade Cycles and American Bank Credit:** Some Aspects of Economic Fluctuations in the United States, 1815-1840. (Doctoral Dissertation, University of Texas, 1958). 1977

Primack, Martin L. **Farm Formed Capital in American Agriculture, 1850-1910.** (Doctoral Dissertation, University of North Carolina, 1963). 1977

Pritchett, Bruce Michael. **A Study of Capital Mobilization, The Life Insurance Industry of the Nineteenth Century.** (Doctoral Dissertation, Purdue University, 1970). Revised Edition. 1977

Prosper, Peter A., Jr. **Concentration and the Rate of Change of Wages in the United States, 1950-1962.** (Doctoral Dissertation, Cornell University 1970). 1977

Schachter, Joseph. **Capital Value and Relative Wage Effects of Immigration into the United States, 1870-1930.** (Doctoral Dissertation, City University of New York, 1969). 1977

Schaefer, Donald Fred. **A Quantitative Description and Analysis of the Growth of the Pennsylvania Anthracite Coal Industry, 1820 to 1865.** (Doctoral Dissertation, University of North Carolina, 1967). 1977

Schmitz, Mark. **Economic Analysis of Antebellum Sugar Plantations in Louisiana.** (Doctoral Dissertation, University of North Carolina, 1974). 1977

Sharpless, John Burk, II. **City Growth in the United States, England and Wales, 1820-1861:** The Effects of Location, Size and Economic Structure on Inter-urban Variations in Demographic Growth. (Doctoral Dissertation, University of Michigan, 1975). 1977

Shields, Roger Elwood. **Economic Growth with Price Deflation, 1873-1896.** (Doctoral Dissertation, University of Virginia, 1969). 1977

Stettler, Henry Louis, III. **Growth and Fluctuations in the Ante-Bellum Textile Industry.** (Doctoral Dissertation, Purdue University, 1970). 1977

Sturm, James Lester. **Investing in the United States, 1798-1893:** Upper Wealth-Holders in a Market Economy. (Doctoral Dissertation, University of Wisconsin, 1969). 1977

Tenenbaum, Marcel. **(A Demographic Analysis of Interstate Labor Growth Rate Differentials;** United States, 1890-1900 to 1940-50. (Doctoral Dissertation, Columbia University, 1969). 1977

Thomas, Robert Paul. **An Analysis of the Pattern of Growth of the Automobile Industry:** 1895-1929. (Doctoral Dissertation, Northwestern University, 1965). 1977

Vickery, William Edward. **The Economics of the Negro Migration 1900-1960.** (Doctoral Dissertation, University of Chicago, 1969). 1977

Waters, Joseph Paul. **Technological Acceleration and the Great Depression.** (Doctoral Dissertation, Cornell University, 1971). 1977

Whartenby, Franklee Gilbert. **Land and Labor Productivity in United States Cotton Production, 1800-1840.** (Doctoral Dissertation, University of North Carolina, 1963). 1977

1975 Publications

Adams, Donald R., Jr. **Wage Rates in Philadelphia, 1790-1830.** (Doctoral Dissertation, University of Pennsylvania, 1967). 1975

Aldrich, Terry Mark. **Rates of Return on Investment in Technical Education in the Ante-Bellum American Economy.** (Doctoral Dissertation, The University of Texas at Austin, 1969). 1975

Anderson, Terry Lee. **The Economic Growth of Seventeenth Century New England:** A Measurement of Regional Income. (Doctoral Dissertation, University of Washington, 1972). 1975

Bean, Richard Nelson. **The British Trans-Atlantic Slave Trade, 1650-1775.** (Doctoral Dissertation, University of Washington, 1971). 1975

Brock, Leslie V. **The Currency of the American Colonies, 1700-1764:** A Study in Colonial Finance and Imperial Relations. (Doctoral Dissertation University of Michigan, 1941). 1975

Ellsworth, Lucius F. **Craft to National Industry in the Nineteenth Century:** A Case Study of the Transformation of the New York State Tanning Industry. (Doctoral Dissertation, University of Delaware, 1971). 1975

Fleisig, Heywood W. **Long Term Capital Flows and the Great Depression:** The Role of the United States, 1927-1933. (Doctoral Dissertation, Yale University, 1969). 1975

Foust, James D. **The Yeoman Farmer and Westward Expansion of U.S. Cotton Production.** (Doctoral Dissertation, University of North Carolina at Chapel Hill, 1968). 1975

Golden, James Reed. **Investment Behavior By United States Railroads, 1870-1914.** (Doctoral Thesis, Harvard University, 1971). 1975

Hill, Peter Jensen. **The Economic Impact of Immigration into the United States.** (Doctoral Dissertation, The University of Chicago, 1970). 1975

Klingaman, David C. **Colonial Virginia's Coastwise and Grain Trade.** (Doctoral Dissertation, University of Virginia, 1967). 1975

Lang, Edith Mae. **The Effects of Net Interregional Migration on Agricultural Income Growth:** The United States, 1850-1860. (Doctoral Thesis, The University of Rochester, 1971). 1975

Lindley, Lester G. **The Constitution Faces Technology:** The Relationship of the National Government to the Telegraph, 1866-1884. (Doctoral Thesis, Rice University, 1971). 1975

Lorant, John H[erman]. **The Role of Capital-Improving Innovations in American Manufacturing During the 1920's.** (Doctoral Thesis, Columbia University, 1966). 1975

Mishkin, David Joel. **The American Colonial Wine Industry:** An Economic Interpretation, Volumes I and II. (Doctoral Thesis, University of Illinois, 1966). 1975

Winkler, Donald R. **The Production of Human Capital:** A Study of Minority Achievement. (Doctoral Dissertation, University of California at Berkeley, 1972). 1977

Oates, Mary J. **The Role of the Cotton Textile Industry in the Economic Development of the American Southeast:** 1900-1940. (Doctoral Dissertation, Yale University, 1969). 1975

Passell, Peter. **Essays in the Economics of Nineteenth Century American Land Policy.** (Doctoral Dissertation, Yale University, 1970). 1975

Pope, Clayne L. **The Impact of the Ante-Bellum Tariff on Income Distribution.** (Doctoral Dissertation, The University of Chicago, 1972). 1975

Poulson, Barry Warren. **Value Added in Manufacturing, Mining, and Agriculture in the American Economy From 1809 To 1839.** (Doctoral Dissertation, The Ohio State University, 1965). 1975

Rockoff, Hugh. **The Free Banking Era: A Re-Examination.** (Doctoral Dissertation, The University of Chicago, 1972). 1975

Schumacher, Max George. **The Northern Farmer and His Markets During the Late Colonial Period.** (Doctoral Dissertation, University of California at Berkeley, 1948). 1975

Seagrave, Charles Edwin. **The Southern Negro Agricultural Worker:** 1850-1870. (Doctoral Dissertation, Stanford University, 1971). 1975

Solmon, Lewis C. **Capital Formation by Expenditures on Formal Education, 1880 and 1890.** (Doctoral Dissertation, The University of Chicago, 1968). 1975

Swan, Dale Evans. **The Structure and Profitability of the Antebellum Rice Industry:** 1859. (Doctoral Dissertation, University of North Carolina at Chapel Hill, 1972). 1975

Sylla, Richard Eugene. **The American Capital Market, 1846-1914:** A Study of the Effects of Public Policy on Economic Development. (Doctoral Thesis, Harvard University, 1968). 1975

Uselding, Paul John. **Studies in the Technological Development of the American Economy During the First Half of the Nineteenth Century.** (Doctoral Dissertation, Northwestern University, 1970). 1975

Walsh, William D[avid]. **The Diffusion of Technological Change in the Pennsylvania Pig Iron Industry, 1850-1870.** (Doctoral Dissertation, Yale University, 1967). 1975

Weiss, Thomas Joseph. **The Service Sector in the United States, 1839 Through 1899.** (Doctoral Thesis, University of North Carolina at Chapel Hill, 1967). 1975

Zevin, Robert Brooke. **The Growth of Manufacturing in Early Nineteenth Century New England.** 1975